The Laboring of Communication

Critical Media Studies

Series Editor
Andrew Calabrese, University of Colorado

This series covers a broad range of critical research and theory about media in the modern world. It includes work about the changing structures of the media, focusing particularly on work about the political and economic forces and social relations which shape and are shaped by media institutions, structural changes in policy formation and enforcement, technological transformations in the means of communication, and the relationships of all these to public and private cultures worldwide. Historical research about the media and intellectual histories pertaining to media research and theory are particularly welcome. Emphasizing the role of social and political theory for informing and shaping research about communications media, Critical Media Studies addresses the politics of media institutions at national, subnational, and transnational levels. The series is also interested in short, synthetic texts on key thinkers and concepts in critical media studies.

Titles in the Series

Governing European Communications: From Unification to Coordination by Maria Michalis
Knowledge Workers in the Information Society edited by Catherine McKercher and Vincent Mosco
The Laboring of Communication: Will Knowledge Workers of the World Unite? byVincent Mosco and Catherine McKercher
Punk Record Labels and the Struggle for Autonomy: The Emergence of DIY by Alan O'Connor

The Laboring of Communication

Will Knowledge Workers of the World Unite?

Vincent Mosco
and Catherine McKercher

LEXINGTON BOOKS

A division of
ROWMAN & LITTLEFIELD PUBLISHERS, INC.
Lanham • Boulder • New York • Toronto • Plymouth, UK

LEXINGTON BOOKS

A division of Rowman & Littlefield Publishers, Inc.
A wholly owned subsidiary of The Rowman & Littlefield Publishing Group, Inc.
4501 Forbes Boulevard, Suite 200
Lanham, MD 20706

Estover Road
Plymouth PL6 7PY
United Kingdom

British Library Cataloguing in Publication Information Available

Library of Congress Cataloging-in-Publication Data

Mosco, Vincent.
The laboring of communication : will knowledge workers of the world unite? / Vincent Mosco And Catherine McKercher.
p. cm. — (Critical media studies)
Includes bibliographical references.
ISBN-13: 978-0-7391-1813-9 (cloth : alk. paper)
ISBN-10: 0-7391-1813-7 (cloth : alk. paper)
eISBN-13: 978-0-7391-2996-8
eISBN-10: 0-7391-2996-1
1. Telecommunication—Employees—Labor unions. 2. Knowledge workers—Labor unions. 3. Mass media—Employees—Labor unions. I. McKercher, Catherine, 1952– II. Title.
HD6976.T24M67 2008
331.88'11384—dc22 2008012491

Printed in the United States of America

∞™ The paper used in this publication meets the minimum requirements of American National Standard for Information Sciences—Permanence of Paper for Printed Library Materials, ANSI/NISO Z39.48-1992.

Contents

Preface and Acknowledgments vii

1 Introduction: New Responses to Bad Times for Organized Labor 1

2 Theorizing Knowledge Labor 23

3 Women and Work: Feminism and Political Economy 53

4 Convergence, Solidarity, and Labor Power: The Dream of One Big Union 81

5 Labor Convergence in the Information Economy 107

6 Beyond Business as Usual: Social Movement Unionism and Information Workers 151

7 Workers in a Changing Global Division of Labor 181

References 223

Index 249

About the Authors 259

Preface and Acknowledgments

This book is the result of a research project, funded by the Social Sciences and Humanities Research Council of Canada, which we launched in 2004. We saw the project as an opportunity to bring together our interests in communication labor, media workers, and trade unions which we had been pursuing separately. We approached the work from the perspective of communication studies, a discipline built around the study of media, messages, and audiences. We have long known that the work that goes into these messages—in other words, knowledge and cultural labor—was a blind spot in the field. Our goal was to help fill it in by undertaking a study of knowledge workers and the organizations that represent them.

Conducting a research project does not, of course, take you out of your own life. As we pursued our work, we were acutely aware of a story in our own community that underscored the importance of knowledge work and the precariousness of knowledge workers.

The story began in 2002 when Maher Arar, an Ottawa software engineer, was coming home from a vacation in Tunisia. His route took him through John F. Kennedy International Airport in New York. There, he was detained by U.S. officials, interrogated about his connection with another Ottawa man suspected of terrorism, shackled, and thrown into jail for close to two weeks. Arar was born in Syria but moved to Canada as a teenager and had been a Canadian citizen since 1991. Instead of returning him home to Canada, U.S. authorities put him on a plane to the Middle East. He wound up in a jail in Syria for more than ten months, where he was tortured and interrogated about links to al Qaeda. He was finally released in October 2003, and on his return to Canada he called for a public inquiry into what had happened to him.

Both while he was in jail in Syria and after his return to Canada, a number of media outlets published stories accusing him of terrorist connections. All citing anonymous sources, these stories were apparently the result of strategic leaks from the Canadian police and security apparatus. Some reporters began to question why the police thought Arar was a terrorist, and how the rendition of a Canadian citizen to a torture chamber in Syria could have happened. A few joined the call for a public inquiry. Others wanted to know why the Royal Canadian Mounted Police (RCMP) considered him a suspect in the first place. In November 2003, Juliet O'Neill, a reporter for the *Ottawa Citizen* and a friend of ours, produced a report that sought to nail down the RCMP case against Arar. Her story relied on a leaked intelligence dossier from an anonymous source that suggested Arar was a trained terrorist and a member of an Ottawa al Qaeda cell. One morning about two months after she wrote that story, ten RCMP officers armed with a search warrant showed up at O'Neill's home. They turned the place upside down. Another group searched her office. O'Neill was pressed to reveal her source, threatened with charges under the Security of Information Act, a piece of legislation enacted following the September 11, 2001, attacks on the United States, and warned she could spend fourteen years in jail unless she gave the police what they wanted. The *Citizen*'s lawyers immediately challenged the warrant and the statute under which it was issued.

The invasion of a reporter's home as part of a criminal investigation into leaks of the dossier raised more questions about the police, and more calls for a public inquiry. A few weeks later the federal government complied. The inquiry took close to two years. It exonerated Arar of any suspicion of terrorism. It found he was the victim of incompetent Canadian police investigators, and also of an orchestrated smear campaign that was designed to fend off an inquiry. The commissioner of the RCMP resigned over the case. Arar himself was awarded $10.5 million from the Canadian government. O'Neill, meanwhile, was still fighting her own case. For close to two and a half years, she lived under the threat of criminal charges. Ultimately, the judge in the case threw out the portion of the security law under which the search warrant for her home had been issued.

So why does this story matter to the authors of a book on communication and knowledge work? Arar is a knowledge worker, a software engineer who came to the attention of the police through bad luck and bad investigative technique. He was also the victim of bad information: the police spotted him having a conversation with another man they suspected of al Qaeda links and assumed the worst. It turns out that the subject of that conversation was utterly benign. The two men were talking about where to buy printer cartridges, tools of their trade. Bad information trailed him to the United States, which chose to send him to Syria rather than return him to Canada. (The United States, by the way, continues to keep him on a terrorist watch list.) He was tortured in an attempt to squeeze information out of him—his alleged knowledge of al Qaeda. When he got home to Canada, he continued to be the target of misinformation. But it was a public inquiry, a full airing of the facts of the case, that exonerated him. O'Neill too is a knowledge worker, in this case a newspaper reporter. She was put through more than two years of court battles and the threat of imprisonment simply for doing her job: presenting information to citizens about the activities of public institutions, all in the public interest. Ironically and painfully, the information in her story turned out to be false, but she had no way of knowing that at the time. When the RCMP invaded her home, the first thing they seized was her computer hard drive, again, a tool of her trade. The search prompted a range of journalists' organizations to rally to her support. One of the first groups to react was her union, the Newspaper Guild sector of the Communication Workers of America, which released a statement pressing the government to reconsider the Security of Information Act under which the raid was launched.

Information matters. Knowledge work is full of complexities. Knowledge workers deal with ideas that can free minds or shackle them. Knowledge workers are individuals, each with his or her own story, but all live their lives within networks of power and control. And often, the best allies of knowledge workers are each other. We dedicate this book to knowledge workers everywhere, but especially to Maher Arar and Juliet O'Neill.

This book drew on the generosity and talents of a wide range of people.

We would like to thank the staff and leaders of a range of trade unions and worker associations who helped us in our work. These include the American Federation of Television and Radio Artists (AFTRA), the Communications, Energy and Paperworkers Union of Canada (CEP), the Communications Workers of America (CWA), the Canadian Media Guild and CWA Canada, the Washington Alliance of Technology Workers (WashTech/CWA), the Telecommunications Workers Union (TWU), and the Screen Actors Guild (SAG).

We are grateful for the research assistance of Enda Brophy, David Lavin, Andrew Stevens, Simon Kiss, and Laura Glithero at Queen's University and of Christopher Bodnar, Kristine Owram, and Amanda Truscott at Carleton University. We are delighted that Carrie Buchanan, also of Carleton University, agreed to take on the task of copyediting this volume, a job she performed with extraordinary skill on our companion book with Lexington, the edited collection *Knowledge Workers in the Information Society* (2007). A special appreciation goes to Brigitte Davlut, whose research assistance on a report Vincent Mosco prepared for a submission to the Canada Industrial Relations Board informs chapter 2. Brigitte died in 2005 after starting a very promising career in media research and intervention. She is missed by many, including those committed to democratic communication.

We would also like to thank the Social Sciences and Humanities Research Council for its generous financial support for the two research grants on knowledge workers that inform this book. We received additional funding from the Canada Research Chairs Program, Queen's University, and Carleton University. We have appreciated the opportunity to present our research in numerous venues in the United States, Canada, Europe, and Asia and benefited enormously from responses and suggestions we gathered in those presentations.

Finally, we are well aware that research in this field is also part of the laboring of communication and culture. Like it, this work is arduous, exhilarating, and painstaking, full of victories and losses. It is a difficult, imperfect, and always unfinished job—a fact we are especially aware of as we start a new round of research. But in many ways and for many reasons, it is, above all, a labor of love.

Chapter One

Introduction: New Responses to Bad Times for Organized Labor

These are not the best of times for organized labor in the United States and Canada. In 2007, 12.1 percent of wage and salary workers in the United States were represented by trade unions. The good news, from the unions' perspective, was that this was up 0.1 percent from the year before. The increase was the first in the United States in a quarter century, and it occurred despite continuing declines in manufacturing jobs (Greenhouse 2008). But however you look at it, an increase of 0.1 percent is a very slight change. And it followed several years of steady decline, from 12.9 percent in 2003 to 12.7 percent in 2004, 12.5 percent in 2005, and 12.0 percent in 2006. In 1983, the first year for which comparable data are available, the rate was 20.1 percent (U.S. Bureau of Labor Statistics 2007, 2008). The situation is somewhat better in Canada, where 29.7 percent of workers were union members in 2007. But in Canada, too, this is significantly lower than the 1981 rate of 38 percent. Canadian trade union membership dropped steadily between 1989 and 1998, but has stabilized at about 30 percent since then (Statistics Canada 2005, 2007).

In both countries, there is a significant gap between union membership rates in the private sector, which formed the backbone of the American Federation of Labor and Congress of Industrial Organizations (AFL-CIO) and its Canadian counterparts for much of the twentieth century, and the public sector. In 2007, the private-sector union membership rate stood at 7.5 percent in the United States and at 17 percent in Canada. The public-sector rate, by contrast, was 35.9 percent in the United States and 71.7 percent in Canada (U.S. Statistics Canada 2007; Bureau of Labor Statistics 2008).

Union membership also varies by age, race, gender, and type of employment. Among demographic groups in the United States, union membership rates were highest in 2007 for African American men (15.8 percent) and lowest for Hispanic women (9.6 percent). More than 16 percent of workers aged fifty-five and older were union members, compared with just 4.8 percent of those aged sixteen to twenty-four. Full-time workers, meanwhile, were more than twice as likely as part-time workers to be union members (U.S. Bureau of Labor Statistics 2008).

We need to place these numbers in historical context. Union membership rates were at these low levels in the 1920s, only to bounce up starting in the 1930s. As late as 1932, an eminent U.S. labor economist, speaking to a meeting of the American Economics Association, reflected on the American Federation of Labor's loss of 40 percent of its members and pronounced that technological change made it nearly impossible for the union movement to regain its earlier strength (Clawson 2003). It is also worth noting that in recent years, the absolute number of union members has remained somewhat more stable than the declining union membership rate might suggest. In the United States, 15.7 million workers belonged to a union in 2007, up 311,000 from the previous year. An additional 1.5 million workers were represented by a union on their main job without being a member of that union (U.S. Bureau of Labor Statistics 2007). In Canada, the number of trade unionists increased by 72,000 to 4.2 million in the first half of 2007. An additional 320,000 workers were covered by a collective agreement (Statistics Canada 2007). This indicates that trade union membership is stable and even growing in some areas, but that this increase has not kept pace with the growth in overall employment.

There are many reasons for the long-term decline in North American trade unionism. While each union or set of workers has its own story, some common threads run through the overall narrative of union decline. Most draw on some combination of technological change, increasing corporate power, the rise of neoliberal governments, and problems internal to trade unions and the labor movement. All have contributed to an overall flattening of wages and benefits for working people. In the United States, median and average wages have stagnated for thirty years, while the availability and quality of employer-subsidized health insurance and pension benefits have eroded substantially. By contrast, the concentration of wealth at the top stratum of society has skyrocketed to levels not seen since the 1920s (Polaski 2007, 1).

Technological change has always had an impact on the organization of the workplace and on worker organizations. The technologies that emerged to create the industrial factory—including steam- or electrically powered machinery and the development of the assembly line—eroded the skills of many craft workers, even as they helped some to become capitalists and managers. They also boosted the development of labor organizations, because the factory system gathered workers together in one place and opened their eyes to the potential power of worker solidarity, including the power to shut down production to press their demands. As the technological development now under way in China powerfully reminds us, the industrial system continues to take hold around the world, creating massive production systems of urban factories as well as turning small-scale agriculture into what Carey McWilliams (1939) would call "factories in the field."

The development of computer systems, starting in the 1960s, significantly altered the process of producing, distributing, and consuming goods and services, including immaterial or information-based goods and services. This has changed the nature of work and the structure, organization, and focus of worker organizations. Much has been made about the shift from industrial to post-industrial production, particularly about the transition from an economy that concentrates on making and moving material goods to one that concentrates on making and moving information and services. In the developed societies of the West and in the urban centers of the developing world where this shift has taken hold, fewer workers are collected in factories. Instead, they gather in offices or work out of their homes. But the question of who controls the workplace remains. Managers have developed a set of measuring and monitoring tools that are both more substantial and more subtle than those used in the factories of an earlier era. Technological convergence—the term refers to the integration of computer and communication technologies—has deepened and expanded opportunities to control work through detailed management of time and space, including virtual space, and through minute surveillance of workplace activity.

In addition, new technologies often offer managers the opportunity to shed workers who perform specific tasks, especially unionized workers. Who needs sound technicians and film technicians when broadcast reporters can digitally record and edit sound and pictures on their own? Why would a newspaper need to purchase new presses, and hire workers to run and maintain them, when it can send pages electronically to a central press operation? Who needs technicians to lay cable in a wireless world? This enhances management's power on another level: the ability to threaten workers with job losses through ramped-up technological obsolescence, or through shipping the work elsewhere. As this book describes, systems sometimes fail and technological convergence does not ensure unlimited managerial control. But technological change has shifted the balance of power in favor of management and made it considerably more difficult for workers to organize and defend their interests. Today's information workers are contained in cubicles, offices, and call-center workstations, or they work alone at home or even out of their cars. They are loosely tied to their fellow workers, often only electronically, and more of their work is monitored. In essence, whether it is an information product or a worker, technological change and technological convergence have enabled a significant, if not revolutionary, expansion in the ability to measure and monitor, package and repackage, resources, assets, and commodities, as well as to shift work from unionized to nonunionized locales. It is therefore not surprising that technological change is considered one of the key factors in the decline of organized labor.

Yet, even when technological convergence is working effectively, there is nothing intrinsic to the computer or telecommunications system that guarantees managerial power, the centralization of control or, as theorists of the labor process would put it, the separation of conception from execution. We have known for some time that information technology systems can be used to advance the widespread distribution of power in the workplace, the dispersal of managerial control, and the continual reskilling or upskilling, not just deskilling, of workers (Zuboff 1988). But examples of these kinds of uses may be best left to the study of deviant cases (Mosco 1994). Especially in those industries that rely heavily on computers and information technology, technological change has been accompanied by the expansion and concentration of corporate power. In other words, technological convergence has driven and been driven by a form of corporate convergence that brings together firms across a wide number of sectors. This is especially prominent in the knowledge- and communication-intensive industries, which can benefit the most from converging technologies.

Throughout the history of capitalism, business has recognized that its job was to expand the value of what economists call the factors of production—land, raw materials, labor, and capital—with technology being used to that end. The latest development in this fundamental recognition is the opportunity to use the systems made possible by computers and telecommunications to expand the value of, and control over, information commodities and labor. As a result, companies are more conscious than ever that they are also in the information business, and that their success depends on drawing value from the information side of the business and from the workers who produce and distribute it. The emphasis on information is found among companies located in the manufacturing, finance, and even resource extraction sectors, but is especially significant in businesses that profit directly from communication and information resources. This helps explain why corporate convergence has occurred at so torrid a pace across the communication and information technology sectors, including in journalism, broadcasting, telecommunications, the Internet, and other new media, as well as in related fields like research, development, and advertising.

Models of corporate convergence are household names in the United States, and they represent a stunning concentration of media and information power. For example,

- Time Warner is a cross-media conglomerate whose divisions include AOL, Time Inc., Time Warner Cable, Home Box Office, New Line Cinema, Turner Broadcasting System (which includes CNN), and Warner Bros. Entertainment (Time Warner 2008).

- AT&T is a leading worldwide provider of communications services to businesses, and the leading U.S. provider of wireless, high-speed Internet access, and local and long distance telephone service. It publishes directories and provides advertising services (AT&T 2008).
- Disney is divided into four major business segments: studio entertainment, which includes Pixar Animation Studios, DisneyToon Studios, Touchstone Pictures, Hollywood Pictures, Miramax Films, Walt Disney Studios Motion Pictures International, and Walt Disney Studios Home Entertainment; parks and resorts; consumer products, including a line of children's books and magazines; and media networks, including ABC television, ESPN Inc., and the Walt Disney Internet Group (Walt Disney Company 2008).
- GE, formerly known as General Electric, has six business areas: GE commercial finance; GE health care; GE industrial, which includes manufacturing; GE infrastructure, which includes products and services in aviation, energy, oil and gas, and transportation; GE money, which consists mainly of banking and credit card services; and NBC Universal, which includes the NBC and Telemundo television networks, film studios, and theme parks (GE 2008).

Other examples of powerful, converged media and information corporations are News Corporation, which owns Fox television along with newspapers, film studios, satellite television, and the book publisher HarperCollins; Microsoft, the world's leading software company, which also owns the MSNBC cable network, the MSN Internet portal, and the Xbox line of game systems; and Google, the Internet search and online advertising company that in 2006 bought YouTube.

Canada has its own corporate giants. Canwest, for example, owns the national Global Television network and a number of specialty television channels. It is also the country's largest newspaper publisher, with one national newspaper, ten major metropolitan dailies, and more than twenty-five smaller daily, weekly, and community papers. It operates broadcast outlets in Australia, the UK, and Turkey, and owns *The New Republic* magazine, published in the United States (Canwest 2008). Bell Canada Enterprises, also known as BCE, includes local, long distance, and wireless phone services; high-speed and wireless Internet access; IP-broadband services; information and communications technology services; and direct-to-home satellite services. Other BCE holdings include Telesat Canada, a pioneer and world leader in satellite operations and systems management, and an interest in CTVglobemedia, which owns the country's leading private television network, its top national daily newspaper, a string of specialty and conventional television channels, and a radio network (Bell Canada Enterprises 2008). Other major players in Canada include Rogers Communications Inc., which operates in wireless telephony, cable television, Internet service provision, consumer magazines, and specialty television; and Quebecor Inc., which specializes in commercial printing, cable services, newspapers, broadcasting, and Internet portals and operates in North America, Europe, Latin America, and India.

Corporate convergence has generally fed into and been fed by technological convergence. More importantly for workers and their organizations, it has significantly increased corporate power. Workers now face global companies whose unprecedented power is made possible in no small part by converging technologies and industries. As with technological convergence, corporate convergence does not always work. For example, the media conglomerate that included the CBS television network and Viacom separated in 2006. At about the same time, Bell Canada Enterprises reduced its corporate stake in CTVglobemedia, which was known from 2000 through 2006 as Bell Globemedia. Nevertheless, the ability of integrated global companies to fend off worker resistance has increased as their financial and technological resources have grown. Companies can finance efforts to stop worker organizations before they are born, or to defang those that manage to appear. Their media units and advertising dollars give businesses new opportunities to influence the perception of workers and their organizations. They have the power to use detailed surveillance to keep track of their workers' activities, including their communication. And when a local situation becomes untenable or when an alternative looks more profitable, they can move to another region or another part of the world. Understanding the extent of corporate power, which is fed by and feeds into technological change, goes a long way toward providing explanations for the decline of labor power.

The obvious reason why the head of the American Economics Association was wrong about the future of trade unions in 1932 was that he could not foresee how, in the midst of the Great Depression, workers would mobilize to defend their interests. But it was not just the economic collapse that fed the revival of trade unionism. Government support for worker organizations, especially enshrining the legal right to organize, helped workers to mobilize from the 1930s through the early 1950s. Whether motivated by the fear of revolution or the need to buy worker support for a wider political agenda, the United States and other governments helped to remedy the imbalance in labor relations so that trade unions actually grew rather than disappeared. Moreover, they grew in areas where jobs were expanding, including in the public service and in the growing fields of media and telecommunications. After the Second World War, in an effort to eliminate radical or progressive tendencies in trade unions that companies could not tolerate, left-leaning unions and their leaders were sacrificed on the altar of the Cold War. Without this once-strong progressive wing, unions did not have the power to successfully defeat attempts to make workers pay mightily to repair the problems of capitalism when its growth engine began to stall in the 1970s.

There are many different ways to characterize the social contract that arose from the ashes of the Great Depression and held sway until the early 1970s, and it had different national inflections (Foster 1988). But in general, the social contract meant that workers would support capitalism, including corporate control over capital and over decisions about management of the workplace. Labor could bargain over some elements of workplace practice, but in general it gave up the goals of worker control and workplace democracy in return for stable employment, full-time jobs (primarily for white men), and modest government support for trade unions. Rising productivity and rising profit levels would be shared with workers through higher wages, pensions, and benefit plans. Public investment in infrastructure and social welfare would provide a social safety net for workers. The government would exercise some oversight and regulation of the economy, reflecting a political consensus of inclusive growth. All parties would ostensibly agree to support a strong national defense complex to defeat communism and to provide a legitimate vehicle for public investment in private enterprise and job creation.

That consensus began to break down in the 1970s. Macroeconomic shocks were partly to blame, including the energy crisis accompanying the rise of a new international oil cartel, and the new and vigorous competition from the economies of Europe and Japan, which had rebuilt the industrial base destroyed by the Second World War. But a larger factor was the rise of neoliberalism, a term that describes new directions for governments and a new set of economic policies and social arrangements. Specifically, neo-liberalism means that governments play a diminished role in public investment and social welfare, in contrast to the decades of Keynesian macroeconomic policies. Corporations are given freer rein to make economic decisions, including decisions that would threaten the stability of employment and the commitment to full-time jobs. Companies would be free to grow in several directions: vertically, to control the entire process of producing, distributing, and marketing goods and services; horizontally, to swallow entire economic sectors; and globally, to benefit from the most efficient use of the factors of production, including labor, wherever it happens to be located. The rise of outsourcing, including offshore outsourcing, facilitated, in some cases, the fragmentation of corporations, with grave consequences for workers who had fought hard to make gains within more centralized parent companies. The newly elected Reagan administration set out to weaken the judicial protections of unions over organizing, collective bargaining, and the right to strike. Perhaps the most chilling symbol of this new attitude was Reagan's 1981 decision to permanently replace striking air traffic controllers. When not openly hostile to workers and citizen movements, governments sought to limit impediments to corporate control, including those initiated by trade unions. Specifically this meant cutting back on worker protections, such as occupational safety and health laws, the social safety net, and other public investments that workers and consumers had fought for over the years in order to limit corporate excess. Fearing a crisis of capital accumulation and seeing an opportunity to tame whatever progressive elements remained in society from earlier labor struggles and from the social movements of the 1960s, governments sought to diminish, if not eliminate, whatever stood in the way of the accumulation process.

The rise of neo-liberalism was not limited to a single country. Led by the United States but supported by Britain, Canada, and others, the inner circle of nations that would come to be called the G8 pressed hard for trade agreements to establish neo-liberalism as the dominant framework for the global political economy. Such a world, citizens were told, offered workers greater material prosperity and more opportunities to compete for wealth. In exchange, workers would have to cope with higher risk and a more precarious life. Not everyone could become Bill Gates, but there would be fewer restrictions on the opportunities to do what it takes to strike it rich.

As with technological change and corporate power, neo-liberalism was not without its failures. The system stalled when nations balked at conforming to a regime that would permit corporate-controlled markets to shape national policies. Social movements coalesced in their own form of convergence to demonstrate for a different form of globalization, one that would be governed by civil society, democratic politics, fair trade, and a commitment to helping the least advantaged first. And after the attacks on New York and Washington on September 11, 2001, the "war on terror"—including the invasion of Afghanistan and the debacle of the war in Iraq—took attention and resources away from the neo-liberal project. The war on terror has, however, had the effect of strengthening the hands of those organizations committed to militarism and to the extension of surveillance to all facets of life, including the workplace.

Nevertheless, it is easy for some to lose sight of just how much neo-liberalism has accomplished, particularly in cementing the power of private corporations over practically all decisions made in the workplace and in many once-public aspects of social and cultural life. Private companies, powerfully protected by the discourse of the marketplace, are now entrusted to advance the public interest. Trade unions, not corporations, are now considered special interests. Supporters of trade unions could once rely on the social contract to provide some room for creating worker organizations. They could also use government agencies to achieve some measure of fairness in the organizing process. Seeing these agencies as impediments to wealth creation, corporations have successfully urged governments to eliminate them or so reduce their powers that labor supporters face a much more difficult path to unionization.

In addition to dealing with the perils created by technological convergence, corporate power, and neo-liberalism, labor organizations are facing internal difficulties that have contributed to their own decline. Trade unions have come in many shapes and sizes, have occupied many different positions on the political spectrum, and have both reflected and resisted mainstream cultural, political, and social forces. But if there is one feature that is common to the movement, it is that trade unions were products of an industrial society. Although they responded to it in many different ways, a manufacturing economy has always been a central reference point for the labor movement. Labor unions represented workers who mined, milled, assembled, built, and packaged material things. They labored in highly structured workplaces, often on assembly lines, in jobs with comparable work routines, in the same communities over many years, and in workplaces comprised mainly of men who were primarily responsible for their family's financial support. And while the unions representing these workers differed in many ways, including in their responses to their common conditions—from the International Typographical Union, which built a model of union democracy, decentralized management, and clean governance, to the International Brotherhood of Teamsters, which built a model of authoritarian and centralized management with notorious consequences—they dealt with a similar set of workplace and wider social challenges.

In the period immediately following the Second World War, the modal workplace began to change from factory to office. Women joined the full-time workforce in unprecedented numbers. Unions made progress in adapting to these changes, but they were slow to shift their organizing efforts to the growing numbers of workers in the information and services sectors. It is understandable that industrial unions fought hard to retain the relative strength they had been able to achieve by resisting the assault of technological change. But this left fewer resources for organizing in the very different world of office work and for bringing in as equal partners the growing numbers of women workers.

Some unions made significant gains, especially among government employees and educators, and some women were included in leadership ranks. But these changes, especially the latter, were limited. Union ranks began their decline. Meanwhile, the leadership of the labor movement found it difficult to accept that unions were in trouble. For example, the head of the AFL-CIO, George Meany, told the organization's 1967 convention that the trade union movement was "a more vital, a more vigorous and a more effective force for progress today than ever before in its history . . . The evidence is there—in abundance . . . It points, not to any decline of force and vigor—but to the beginning of a new and challenging period of growth." In a 1972 interview,

Meany dismissed concerns about the decline in union density, saying: "To me, it doesn't mean a thing. I have no concern about it, because the history of the trade union movement has shown that when organized workers were a very, very tiny percentage of the workforce, they still accomplished and did things that were important for the entire workforce" (quoted in Dark 1999).

Shrinking numbers and shrinking resources contributed a further problem that ultimately led to a split in the major labor federation in the United States. Unions have always struggled over the allocation of resources for three different tasks: organizing to bring new workers into union ranks; collective bargaining over wages and working conditions; and political activity such as lobbying, working to elect prolabor candidates for public office, contributing financially to political or social campaigns, and so on. Disagreements over the best mix among the three are part of the process of setting priorities for unions and for labor federations. But these differences grew more significant over time as a changing workplace and a changing composition of the workforce made increasing demands on major labor organizations. Significant splits arose as some of the older, industrial unions focused on keeping contract gains and on working to support labor-friendly candidates for public office. By contrast, some of the newer and more aggressive service unions fought to devote more energy and resources to organizing. These divisions came to a head in September 2005 when seven unions, representing 6 million workers, broke away from the AFL-CIO to form a new labor federation called Change to Win. Claiming that the AFL-CIO was too traditional in its makeup and wasn't paying enough attention to organizing at the grass roots, Change to Win sought to build on the success of its largest member, the Service Employees International Union, in organizing new groups of workers and new types of workplaces. At a time when business mergers were accelerating and business lobbies were tightening their influence on Washington, the major labor coalition broke apart. Given the declining power of unions, it should come as no surprise that when the Democrats took control of the U.S. House of Representatives and Senate in 2006, legislation to ease the process of organizing a union was not one of their major priorities.

If this book simply was about organized labor's decline as a result of technological change, increasing corporate power, the rise of neo-liberal governments, and internal divisions among unions, there would be no need to proceed any further. Others have told this story in the popular media and in academic accounts. It is an important story, but it is not the only one that can be told about organized labor. So while the narrative of labor's decline, of necessity, informs this book, we seek to tell other stories. These suggest how labor is organizing in new ways and responding—yes, often defensively, but also creatively—to technological change, rising corporate power, and neo-liberalism.

In one very prominent pattern, organized labor has undertaken its own form of convergence, responding to technological and corporate convergence in the knowledge economy by bringing together workers once divided by technological, craft, and industry barriers. Journalists in print and broadcasting have joined in the same union with technical workers who operate cameras and soundboards; media workers have united with unionized telephone operators and cable installers; telecommunications workers have come together with call-center workers, software creators, and other workers in the knowledge, communication, and service trades to produce trade unions that span their converging industries. In one case, Canada's Communications, Energy and Paperworkers Union, the converged union even includes the people who produce the newsprint for the print publishing industry. These converged unions encompass highly paid evening news anchors and freelancers who live precariously, one story at a time. They operate nationally, binationally, and globally. Unions are using this convergence strategy to win victories with individual employers, as well as to build ties across the once-significant divides that separated sectors, crafts, skills, statuses, and cultures. These unions are creating alliances that overcome enormous international barriers, including the daunting line that divides knowledge workers in rich countries like the United States, who are increasingly the victims of outsourcing, from the growing class of knowledge workers in poorer nations like India, who have benefited from this process. In one sense, these unions are reviving the historic vision of "one big union," which imagined bringing together workers across all of the great divisions that have traditionally constrained opportunities for resistance and the pursuit of a worker agenda. The contemporary vision is not so ambitious; it is concentrated in the communication, media, knowledge, and related service sectors. Observing the rise of an increasingly integrated electronic information and entertainment services industry, led by integrated corporate conglomerates, unions are responding with their own versions of convergence for the information age.

Another related story describes how new forms of worker organizations, again active in the knowledge economy, have provided important representation for those employed in the hardest-to-organize sectors, such as the computer industry. Sometimes with the help of newly converged unions, the workers who produce software, write technical support manuals, or patch security breaches in programs have come together in the challenging task of providing collective representation at Microsoft or IBM. It is notoriously difficult to organize a union at any private employer today, but it is especially challenging in the high-tech industry. Indeed, some might wonder why the beneficiaries of high pay and stock options would even feel the need to think about a union—and understandably so. But in reality the computer industry is deeply divided among three kinds of workers. At the bottom are those who manufacture or assemble hardware, including processors, screens, and peripherals. Increasingly stocked by workers in China and the rest of Asia and by immigrant women in the United States, they earn little and suffer the worst occupational safety and health problems. At the top of the hierarchy are full-time permanent workers who have enjoyed high salaries and rich benefits, including, in some cases, stock options. They often put in long hours at work that is simultaneously creative and grindingly repetitive. Despite the popular perception, their working days are not filled with vast stretches of foosball and video games. But they enjoy far more benefits than the third type of worker, the similarly skilled but temporarily employed workers who toil alongside them. This group includes people who also put in long hours doing the same kinds of work as permanent workers, but are paid an hourly wage and have few of the benefits enjoyed by their more fortunate colleagues on the permanent staff. In addition to these significant material distinctions, status divisions abound.

These workers are often made to wear different-colored identification badges, and they are shut out of events designed to build corporate culture. These permanent temporary workers, known as permatemps, are increasingly the models of labor in the contemporary workplace, especially in the communication and information industries. They fill the ranks of the new "precariat," or precarious workers, so named because they enjoy few of the benefits and more of the risks than their full-time counterparts. One of the interesting stories in recent labor history describes how the hardware manufacturers and permatemp knowledge workers are creating organizations to defend their interests. Neither group is likely to form a traditional trade union. But many have taken to new forms of labor organization that meld some aspects of traditional trade unionism with social activism. These include worker associations that provide concrete benefits to members but are also involved in wider political issues such as the fight against outsourcing, or social movement organizations that help members primarily through participation in broad political and cultural movements that see labor issues as human rights issues. Often with the assistance of traditional unions, these groups may provide legal and health services, run courses, offer counseling, lobby locally and nationally, and connect their members to groups engaged in social activism on issues like environmental protection or immigrants' rights. They have enjoyed some success, including a substantial court victory by the Washington Alliance of Technology Workers (WashTech) against Microsoft that challenged the very notion of the permanent temporary worker. As a result, it is important to consider these as green shoots on the otherwise bleak landscape for organized labor in North America.

Admittedly, it is uncertain whether the stories of convergence among major unions, or of new forms of organizing in unlikely places, represent a new dawn for the labor movement or its last defensive gasps. We are inclined to lean to the former, with several caveats that the book explores. In order to make reasoned judgments about the future, it is important to understand what is taking place today. No story about labor, especially in the growing knowledge sector, is complete without this understanding. And no study of the growing knowledge sector, we contend, is complete without an understanding of the labor that goes into it. In essence, having spent a great deal of time and energy wondering about what will be the next new thing, the next technological gadget or "killer app," it is time for communication scholars to pose an old question for a (more or less) new age: Will knowledge workers of the world unite?

There is, of course, no definitive answer to this question, but simply to ask it and to explore a set of possible responses goes against the current of the leading stream of scholarship on the subject of knowledge work. Specifically, much of the research on knowledge labor draws from theories of postindustrialism and informational capitalism that present the characteristics of a new social form or a variation on an older form. This work contributes a great deal to debates about knowledge workers, but it comes with a clear tendency. Whatever other differences may exist in this stream of theory, it presents society as the product of structural forces—technologies, corporate decisions, and government policies—that most workers cannot lead or even change in any fundamental way. Asking whether knowledge workers of the world will unite presumes that workers are active agents in the construction of this society, not just passive recipients of its impacts. And in posing that seemingly provocative question, we are not alone: the organizations that represent workers, in North America and elsewhere, are asking the same question. One of this book's departures is a focus on how workers actively organize their world rather than simply adjust to its exigencies. We make no claim to offering the full range of worker activity. Rather, we have chosen to focus on efforts to explore new forms of organization in the knowledge-based society.

Overview of the Book

Immediately following this introductory chapter, the book presents two conceptual chapters that examine different ways to think about knowledge workers and the world they occupy.

Chapter 2 provides a set of definitions of knowledge labor, starting with a narrow focus on the creative work of content production, expanding the definition to include those who work on packaging and distributing information, and expanding it still further to encompass all those employed throughout the chain of value in the knowledge sector, including those who work on the production of hardware. In an effort to shed light on debates about the nature of the social world occupied by knowledge workers, the chapter then takes up different ways of thinking about the information society, ranging from those who see it as a wrinkle in the development of capitalism to those who view it as a substantial disjunction in capitalism that is likely to lead to a transformation to a new social order.

The chapter proceeds to examine four periods in the development of theory and research on knowledge workers. The first suggests that work in this area is not new, as is demonstrated in the writing of Saint-Simon in the eighteenth century and Marx and Babbage in the nineteenth. Next comes a discussion of the first rush of systematic research on the subject, starting in the late 1940s and continuing into the 1960s. The perception that Western societies, led by the United States, were experiencing an occupational shift from manufacturing to information and service jobs was tested by Fourastié, Machlup, and, in greatest depth, Porat. Each of these documented the shift, began the process of conceptualizing it, and prepared the way for more serious theorizing to come. A trio of books, published in 1973–1974, mapped a significant portion of the territory on the place of knowledge workers in an information society. Daniel Bell's *The Coming of Post-Industrial Society* (1973) described a shift in power from inherited wealth and control over manufacturing to a meritocracy based on education and skill in the production and distribution of information. Herbert Schiller's *The Mind Managers* (1973) was an implicit response to Bell, arguing that the concentration of business power over the production of media and information both deepened the power of capital and challenged the free exchange of information and ideas that is central to a democratic society. Harry Braverman's *Labor and Monopoly Capital* (1974) focused the critique at the point of production, where the shift to jobs based in information, communication, and services had the effect of deepening managerial control over the workplace. This accelerated the split between the conception of a task and its execution, concentrating the power of conception in a managerial class and deskilling the mass of jobs.

The final layer of analysis addresses contemporary research that has explored new areas, questioned earlier analyses, and deepened their research. For example, Manuel Castells (2000, 2003) has built on Bell's edifice with an analysis of informational capitalism. Ursula Huws (2003) has sharpened Braverman's critique of the information-based labor process and extended his work by taking the gendering of labor as a central theoretical tool. Nick Dyer-Witheford (1999) has deepened Schiller's research with a general critique of the capitalist political economy of information, but broadened it as well by examining the centrality of labor and new technology as forces for democratic social change. Others, like David Naguib Pellow and Lisa Sun-hee Park (2002), situate knowledge work in a transformed international division of labor in which business takes advantage of cheap labor at the base of the information society food chain. The chapter concludes by documenting how the book builds on this theoretical foundation.

The knowledge workplace, including those organizations taking a leading role in representing workers, is increasingly female. But the literature on knowledge labor, with important exceptions, has not dealt substantially with this development. Chapter 3 takes a step toward remedying this by demonstrating how to build bridges between feminist and political economic theory to advance our understanding of knowledge labor. Feminist scholars have long recognized the conflicted nature of women's relationship to both work and home, including competing demands for women's time and attention, and competing desires for professional and personal satisfaction. This chapter examines four interrelated perspectives on a feminist political economy of labor and communication, organized around responses to the converging worlds of work and home. These perspectives are not mutually exclusive; indeed, a primary purpose of the chapter is to demonstrate how each can enrich and inform the other. All four take seriously the need to build connections between political economic and feminist research, and all four recognize the importance of the communication, information, and knowledge arena as a substantive base for exploring that relationship.

The chapter begins with a discussion of recent scholarship on women's unpaid work in the home, demonstrating that much of this labor is tied up with buying and maintaining an ever-expanding list of digital devices. This perspective suggests that capitalism commodifies the domestic environment by promoting the view that the value of what takes place within the home depends on the value of the tools, toys, and products that fill it. As the need for wage labor to maintain the home grows, women whose household labor goes unrewarded in the wage system are finding ways to earn money. The chapter then takes up some of the issues surrounding wage labor carried out in the home, focusing on the computer industry in California's Silicon Valley. Here, homework is not unpaid domestic labor; it is the wage labor of women with few alternatives to turning their homes into extensions of factories or call centers. These women, many of them immigrants, work at home as "independent contractors," producing or assembling computer components. This means they work with toxic chemicals, exposing themselves and their families to environmental hazards. One of the valley's dirty little secrets is the use of these women to carry out essential work. The third section of the chapter deals with scholarship on the erosion of the separate categories of work and home.

Drawing mainly on the work of the autonomous Marxists, this scholarship concentrates on the emancipatory potential of new technologies and the accompanying rise of a knowledge-worker class. In this point of view, the home and the traditional workplace both become places of work and play, mutually constituted and a fundamental challenge to capitalist efforts to reduce all activity to wage labor. The fourth section takes up the question of how women can build a home in the house of labor, including in traditional trade unions, social movement unions, and labor federations. Membership in a trade union can provide significant economic benefits to women workers. Nonetheless, the labor federations are still, to a large extent, the domain of men. The chapter maintains that it is not possible to fully understand the relationship of gender to the workplace without understanding the relationship of the workplace to the home. The home, after all, has been the center of women's work until relatively recently. And even when the work that women do moves out of the home with the development of wage labor, it still bears a distinct relationship to the domestic sphere.

With this theoretical grounding, the book proceeds to consider the historical origins of new developments in trade unionism. One of the critical issues facing worker organizations is *how* to build labor power. Chapter 4 explores the strengths and limitations of two different approaches to the problem: asserting power based through craft control and asserting power through mass or class-based movements. It begins with the case of the International Typographical Union (ITU), the dominant union in the newspaper business for much of the twentieth century and one of the strongest unions for its size in North America. The ITU was the original "one big union" in communication, at one point representing all stages of the work of putting ideas on paper. Its strengths rested on its ability to maintain jurisdiction over the work, and to grab and hold the loyalty of its members. The chapter describes how the union prevailed over the introduction of the linotype machine in the nineteenth century. But it shows that at about the same time, the union made a grave miscalculation: it failed to keep in the fold the other workers who toiled alongside the linotype operators, prompting them to break away and form craft unions of their own. This hampered its prospects for controlling the next wave of technological innovations, in the middle of the twentieth century, and contributed ultimately to the demise of ITU craft control. The chapter then takes up some of the early North American efforts at building broadly class-based unions, touching on the Knights of Labor, the Industrial Workers of the World (IWW), and the One Big Union. These groups contained internal contradictions that contributed to their downfall, but their most potent enemies were the forces of capital and government, which sought to rid labor of its most radical elements. Contemporary convergent unions trace their roots to both craft unionism and to the movements to create one big union for the working class. The chapter concludes with a discussion of the lessons these unions might draw from the experiences of both craft and mass solidarity movements as they seek ways of organizing in a globalizing economy. One of these is the idea that craft identity, however it may be defined, matters to workers and can be a tool for building stronger unions. A second is the idea that size does not necessarily equal power. Finally, one of the key lessons from all the earlier attempts at building labor power is a reminder that the state and the forces of capital are not idle observers of the process of assembling labor power. As a result, labor unions have to look outward, keeping an eye on what their employers are doing, and what they are likely to do next.

Chapter 5 takes up the effort to build convergent unions across the communication and information industries of North America. It begins by sketching the rough outlines of labor convergence in the communication and information sector in the United States and Canada, identifying several forms of convergence. These include mergers among national unions, bargaining-unit consolidation within individual workplaces, and the activities of national or international labor federations. The chapter then takes up three case studies in trade union convergence. The first looks at the Communications Workers of America (CWA), perhaps the best example in North America of a union with a focused convergence strategy. Among its successes was achieving bargaining-unit consolidation among English-language workers at the Canadian Broadcasting Corporation. The strength of this new "one big union" was tested in 2005, when the employer locked out the workforce in a dispute that centered on the CBC's desire to move toward a contingent workforce. Ultimately, the corporation had to back away from this goal. The bargaining-unit consolidation at the CBC has turned out to be a striking example of how a single union is a better voice for workers of all sorts, and where diversity among members contributes to strength and single-minded determination among the leaders. Nonetheless, pressures from management to reduce the unionized labor force by outsourcing and cutting jobs will continue.

The point of chapter 5 is to document the strengths and limitations of trade union convergence for workers in the knowledge sector. The other two case studies it takes up show what happens when unions choose not to converge, or to pursue a partial convergence strategy. One of these cases takes us to Hollywood, where the film and video business is massively converged but where workers are organized in a set of narrow unions and guilds. This case study examines attempts to bring together some of these units, focusing on the three attempts to unite the Screen Actors Guild (SAG) and the American Federation of Television and Radio Artists (AFTRA). It considers why none of these efforts have succeeded and offers lessons on the negative fallout that may result from the failure of convergence—specifically, competition between the unions in which the employers, not the workers, are the ultimate winners.

In many cases of labor convergence, smaller unions tend to become stronger by joining a larger unit with more resources and a stronger organizational base. But what happens when a small but strong union with a radical tradition decides to go it alone against an integrated conglomerate? The final case study in chapter 5 considers Canada's Telecommunications Workers Union (TWU), which brought a history of militancy to its recent conflict with Telus, Canada's second-largest telecommunications provider. Partly to protect the gains it won over years of battles with Telus's predecessor, BC Tel, including a veto over technological change, the TWU fought hard against Telus, a more potent and better integrated power than BC Tel and one with close ties to neo-liberal governments at the federal and provincial levels. After a long lockout and strike, the union ultimately signed an agreement in which it gave up some of its extraordinary powers and accepted what has become a standard agreement in the telecommunications industry. The case study addresses the reasons for this outcome, including the importance of choices made about convergence. The chapter concludes that labor union convergence offers no guarantees of success at the bargaining table. Nonetheless, it can be a powerful force for working people, and it has the potential to allow unions to "bite back" at companies that have used technological and industry convergence to chew up their unionized workforce.

One of the more prominent stories in coverage of organized labor today is the difficulty workers face in drives to create unions, especially in the United States. Chapter 6 begins with an analysis of two leading ideas on how to reinvigorate the labor movement in North America. One, known as value-added unionism, concentrates on transforming unions into organizations that offer positive gains to employers, thereby making themselves more valuable in the workplace. The second, known as social movement unionism, advocates reinvigorating trade unionism from the grass roots up, by overthrowing moribund bureaucracies and connecting unions with other worker organizations and community groups. The chapter then looks at examples of new forms of organizing in the communication and information sector that are inspired by, or connected to, the social movement model. One of these is the Washington Alliance of Technology Workers (WashTech), which originated among contract workers at Microsoft and has been active in bringing national attention to problems associated with both offshore outsourcing of knowledge labor and bringing low-wage information technology workers into the United States. Others include a range of groups in Silicon Valley that provide assistance and support to workers in the hardware production side of the computer business. In addition to assessing their activities, the chapter examines their relationships to other unions—shortly after its inception WashTech became a local of the CWA—and to social movement and environmental justice organizations. This provides an opportunity to consider the viability and future prospects for such organizations. The chapter concludes that we are likely to see expansion in social movement unionism in the coming years.

The final chapter takes us into the international arena, where the labor movement faces what is arguably its most daunting challenge. Globalization is bringing about massive changes in the international division of labor as companies make use of communication and information technology to expand their operations wherever they can find the most cost-efficient means of production. This is particularly the case in the knowledge sector, which is less burdened than others by the material weight of its output. The observable gains of convergence among labor organizations in the developed world are fiercely challenged when companies outsource the jobs that these organizations have fought hard to organize, maintain, and enrich with promises of employment security and other benefits. This chapter addresses the nature of outsourcing in the knowledge industries, including the antinomies that complicate arguments on both sides of the debate. Outsourcing is not simply a matter of transferring jobs from developed to developing world. Jobs go everywhere, and precisely where they go (or for how long) depends on a range of factors.

The chapter's central focus, however, is on what labor is doing about it, taking up two case studies on approaches to combating offshore outsourcing. The first deals with how the Hollywood unions have fought against so-called runaway production, a term that refers to the filming of American stories intended for American audiences in other parts of the world, such as Canada or eastern Europe. The case describes how the unions got the issue on the national agenda in the United States and convinced a number of state governments in that country to subsidize film production. The fight has met with some considerable success. At heart, though, the effort to combat runaway production is aimed at protecting jobs for workers in one country, which raises questions about the long-term consequences of efforts that pit worker against worker. The next case takes up the idea of building coalitions across the divides that separate workers in rich and poor countries. It focuses on the Union Network International (UNI), a federation of unions primarily in the media, information, communication, and services industries. Since its founding in 2000, UNI has been active on several fronts, promoting an ambitious agenda that covers global organizing, global framework agreements, global objectives for decent work and decent rates of pay, and minimum standards for health and safety. It also works for collective bargaining rights for all, global solidarity action, global policy development and research, and global representation. UNI has become one of the central nodes of labor opposition to offshore outsourcing, forging ties between workers in developed countries who are losing jobs and workers in developing countries who are the recipients of offshored jobs. These connections are an important consideration in addressing the question of whether knowledge workers of the world will, indeed, unite.

History of the Project

For some time, the authors of this book have studied labor issues in diverse segments of the U.S. and Canadian communications industry. Vincent Mosco's research on labor in the communications industry dates back to a 1983 book with Janet Wasko that addressed theories of labor in a postindustrial society, the communication industry labor process, and media representations of workers (Mosco and Wasko 1983). He has also carried out a national study of Canadian telecommunications workers and provided expert reports on labor and convergence in telecommunications to assist with three cases before the Canada Industrial Relations Board (Mosco 2002; Mosco, Zureik, and Lochhead 1989). Catherine McKercher has written about the impact of new technology on journalism. This led to a book on how trade unions representing newspaper workers in the United States and Canada have responded to technological convergence, corporate concentration, and changes in the labor process (McKercher 2002).

This immediate project began in 2004 with a grant from the Canadian Social Science and Humanities Research Council to explore trade unions and convergence in the North American communications industry. As scholars who have studied the media, communication, and information technology over the years, we have been well aware that the field has paid inadequate attention to labor. Research in communication studies tends to cluster around the exploration of three interconnected topics: media, messages, and audiences. Scholars who focus on media tend to examine questions of power and control, including media ownership and the social, political, and economic relations that are at play in the construction of messages and of audiences. Research on messages tends to look at content, ranging from news to propaganda to advertising, and at the discursive and technological forms these messages take. Studies concentrating on audiences tend to examine how individuals or groups receive, make sense of, understand, act on, ignore, or incorporate messages into daily life.

And yet it is clear that intellective and physical labor are required to create messages, as well as the technologies used to disseminate them. Receiving and acting on messages also requires labor. Media scholars generally recognize this, but they rarely address the various forms this laboring takes. Similarly, the organizations that represent media and information workers, and the presentation of labor in the media, have also received relatively little attention. It is probably accurate to conclude that if, as the media theorist Dallas Smythe (1977) famously remarked, communication is the blind spot of Western Marxism, then labor remains a blind spot of communication studies.

The first fruit of our project was a special double issue of the *Canadian Journal of Communication* (2006) whose theme, "the labouring of communication," is also the title of this book. In choosing the theme and the title, we were inspired by Michael Denning's masterful cultural history of the United States in the middle years of the twentieth century, *The Cultural Front* (1996). Denning uses the term "the laboring of culture" to sum up a number of interrelated themes, all of which put labor at the forefront of the cultural struggles of the era (1996, xvi–xvii). The period from the Great Depression to the 1950s saw the word "labor" join "work," "industry," and "toil" as key words in the vocabulary of cultural workers. As a result, the language itself was "labored." Those years also saw the increased influence on—and participation of—working-class Americans in the arts, media, and culture. This was largely the result of a rapid expansion of mass education and mass entertainment, as children of immigrants and working-class families grew up to become artists and employees of the greatly expanding cultural industries and as American workers became the primary audience for those industries. Denning also uses the phrase "laboring of culture" to refer to the new visibility of the labor that went into cultural production. He contends that one of the central narratives of the era was the organization of cultural workers into labor unions. Those swelling the ranks included teachers, newspaper reporters, motion picture actors, and radio stars, as well as the workers whose technical expertise ensured that people could watch the movies, listen to the music, and communicate with each other about how to consume the products of the cultural industries. The laboring of culture is also a reminder that culturally and politically, the middle years of the last century were characterized by working people embracing social democracy, or the fullest possible participation in economic, political, and cultural decision making. Finally, Denning contends that the laboring of culture connotes one of the earliest conceptions of labor: the work that leads to a birth. The laboring of culture, therefore, entailed work and toil, and had its successes and failures. As he puts it: "To labor is to plod, to be hampered, to pitch and roll in a storm. In all these senses, the cultural front was a laboring, an incomplete and unfinished struggle to rework American culture, with hesitations, pauses, defeats and failures" (1996, xvii).

The articles in the special issue of the journal were written by scholars working from a variety of perspectives and in a variety of countries. The common factor was that regardless of disciplinary or national boundaries, they put labor at the forefront.

That emphasis continued in the second major product of our project, *Knowledge Workers in the Information Society* (McKercher and Mosco 2007). This coedited book was designed as a companion reader to the present volume. In a set of interlinked chapters, it addresses the changing nature of work, of workers, and of their organizations in the media, information, and knowledge industries. These include newspaper journalists, broadcasters, librarians, film and animation workers, and government workers providing social assistance using new technologies, as well as employees in the telecommunication and high-tech sectors. The chapters take up the history and theory of knowledge work and provide a set of empirical case studies on the experience of knowledge workers in societies increasingly characterized by the production, distribution, and consumption of information. The book is multidisciplinary and international. With authors trained in anthropology, communication studies, history, film studies, industrial relations, journalism, library and information science, media studies, sociology, and women's studies, the book addresses knowledge workers from several interconnected perspectives. Furthermore, with contributions from authors based in the United States, Canada, Spain, the UK, Australia, and Taiwan, as well as authors who have had extensive research experience in China, India, and other parts of South Asia, the book examines knowledge workers from a multinational and, most importantly, a genuinely global perspective.

This book offers the results of our own research on the subject, with a specific emphasis on how communication and information workers in the United States and Canada are organizing their response to changes in the workplace, nationally and internationally. Between them, we hope that this volume and *Knowledge Workers in the Information Society* will help to bring labor issues in from the fringes of the discipline—to make labor and those who perform it part of the common vocabulary of communications scholarship.

Chapter Two

Theorizing Knowledge Labor

The study of knowledge labor and the information society has raised important questions for academics and policymakers. Because these groups have a significant impact on research and intervention, some of the most fundamental questions they address have to do with how to define and make use of these terms. There is extensive debate on this issue and, as we might expect, this has resulted in a multiplicity of definitions. The term "knowledge worker," for example, has been applied to groups as diverse as university professors, software designers, journalists, call-center workers, and United Parcel Service truck drivers. The term "information society," meanwhile, has been conflated with, or replaced by, concepts like postindustrial society, network society, or digital capitalism. The range of definitions grows out of a long-standing debate over how to conceptualize knowledge labor, including how knowledge labor relates to capitalism. In recent years, the debate has been complicated by the phenomenon known as convergence, which refers to the bringing together of media forms, institutions, and technologies. As with the other terms, there are conflicting conceptions of convergence, leading to conflicting questions over its impact. For our purposes, the critical question is how convergence relates to knowledge labor and to the trade unions and other organizations that represent knowledge workers.

This chapter seeks to provide some guidance on how to think about all these concepts, and how they relate to communication workers and worker organizations. It begins by examining a range of definitions for knowledge labor and the information society. It then examines different perspectives developed since the end of the Second World War on how to conceptualize knowledge labor. Next, it takes up the many manifestations of convergence, including labor convergence. It concludes by considering how four key concepts enable us to enlarge our vision of knowledge workers in the information society: technology, communications, democracy, and citizenship.

Defining Knowledge Work and the Information Society

Given the extensive debate over the meaning of the terms "knowledge work" and "information society," we feel it is most useful to provide a range of definitions for each rather than to imagine, and impose, a single, ostensibly correct meaning.

At one end of the continuum, knowledge work has been defined narrowly, applying only to labor that is directly creative. For example, Florida (2002) limits knowledge work only to those jobs that directly manipulate symbols to create an original knowledge product, or add obvious value to an existing one. In this respect Florida draws on a tradition of thinking dating back to the work of Peter Drucker (1959, 1993), who started writing about a new class of creative workers in the 1950s, and to Robert Reich (1991), who defined knowledge workers as symbolic analysts. According to this view, knowledge workers are people like writers, artists, web-page designers, software engineers, and university professors. Specifically, Florida envisions two types of knowledge workers. One group is a supercreative core of people who produce "new forms or designs that are readily transferable and widely used" (2002, 68). They work in science and engineering, education, computer science and mathematics, and the arts, design, and entertainment industries (2002, 74). Just beneath this core is a creative class of people whose work requires creative problem solving, addressing concrete problems by drawing on complex bodies of knowledge. They work in high tech, business management, financial services, legal services, and health care, as well as in the cultural industries encompassing the media and the arts. While not quite at the level of those who create new forms, the members of this creative class are highly educated people whose major value lies in their ability to translate human capital into creative work. In addition, for Florida and Reich, the system is dynamic. This means that individual members of the lower ranks can develop the skills to move into the creative or symbolic analyst class, or an entire occupation—professional secretaries for example—can move up into the creative category. Although this appears to be a narrow segment of the workforce, Florida counts about 10 percent of the U.S. labor force in the supercreative core, with an additional 30 percent or so in the creative category.

A more expansive definition of knowledge work encompasses the labor of those who handle, distribute, and convey information and knowledge. This includes secondary and elementary school teachers, journalists, and librarians at the more knowledge-intense end of the category and media technicians and postal workers at the other end. (See, for example, Day 2007; Deuze 2007; Kumar 2007a; Martin 2007; Tracy and Hayashi 2007.) The argument in favor of considering these to be knowledge occupations is that an increasing amount of the work involves making use of information or information technology to efficiently and effectively deliver an information product. If both the tools used by the worker and the product the worker delivers are based on information, then it is difficult to argue against considering that worker to be an information worker.

Finally, at the other end of the continuum, we find the most expansive definition of the knowledge worker: anyone in the chain of producing and distributing knowledge products. In this view, the low-wage women workers in Silicon Valley and abroad who manufacture and assemble cables and electronic components are knowledge workers because they are an integral part of the value chain that results in the manufacturing of the central engine of knowledge production: the computer (Pellow and Park 2002; Smith, Sonnenfeld, and Pellow 2006). Similarly, call-center workers, who sell communication products and services, would also fall within this broad definition of knowledge work because they are central to marketing information and because they make use of the products of communication technology to carry out their work. Indeed, some would add that the management and control of their work would be far more difficult were it not for the advanced surveillance technologies made possible by developments in communication and information technology (Head 2003).

Another way of bounding the term is to look beyond external criteria and examine the subjective experiences of the workers themselves, including the ways they choose to organize or structure their trade unions. Here we see some evidence to support a broader conception than those offered by Florida or Reich. Unions like the Communications Workers of America (CWA) and its Canadian counterpart, the Communications, Energy and Paperworkers Union of Canada (CEP), have brought together a diverse range of workers under the same umbrella—journalists and telephone operators, translators and customer service providers, health-care workers and printers, the people who write and broadcast the news and those who work the cameras and soundboards to bring it to viewers and listeners, and so on. These two unions have had some success in mobilizing their diverse memberships in effective labor actions (Mosco and McKercher 2006). However, the mixed pattern of labor convergence across the communication sector in North America, including the failure of some creative unions to converge, leaves open the question of how trade unions themselves bound the term "knowledge worker" (McKercher and Mosco 2007).

Debates about the scope of how to conceptualize knowledge work are also debates about social hierarchies. Since much of the research on knowledge work tends to see inclusion in that category as a positive development for the workers who make the move, the type of definition one accepts is also an implicit decision about who and how many are privileged. Limiting the definition to Florida's creative class, or even to his supercreative core, gives greater weight to the privilege of being a knowledge worker than accepting a broader definition that covers different types of workers, including many, such as call-center employees, about whom it is more difficult to justify the privileged label. A creative class opens the door to one type of political change, such as a shift in power to knowledge creators. A more heterogeneous vision of the knowledge-work category points to another type of politics, one predicated on questions about whether knowledge workers can unite across occupational or national boundaries, whether they can maintain their new-found solidarity, and what they should do with it. Moreover, by limiting the concept to occupations that are centrally based on knowledge work or that are directly involved in chains of symbol production, it is easy to miss the point that all work is, in some respects, knowledge work, and that many knowledge workers in what are typically described as working-class jobs are ignored in the hoopla over the information economy (Livingstone and Sawchuk 2004). The "hidden knowledge" that is central to many jobs is also important in the politics of organized labor, and a relatively inclusive definition of knowledge work makes it easier to account for it.

Another significant division in the concept of knowledge work holds important but different political implications. Think of the continuum we've just described as operating along a vertical axis, with the narrowest definition of knowledge workers (i.e., directly creative workers) at the top and the broadest definition (i.e., those who build hardware or are part of the chain of production) at the bottom. Knowledge workers can also be divided along a horizontal axis, falling into a pure content category at one end of the continuum and a pure technical category at the other. Content and technical groups both include creative workers in Florida's meaning of the term, but many other types of workers as well. The content category encompasses artists, entertainers, teachers, journalists, musicians, and others who might also be called cultural workers. The technical category covers software designers, biomedical engineers, audiovisual technicians, and those whose creative contribution is the construction of code, the design of technologies, and the production of signals. The labor of this category makes possible the labor of cultural and other producers of what is typically viewed as content. Examined this way, the difficulties of calling all these people knowledge workers become readily apparent. What do engineers and writers have in common? Journalists and telecommunications specialists? Musicians and cancer scientists? Regardless of whether the concept of knowledge worker is defined broadly or narrowly along the vertical axis, we are still left with the divide between those who focus on creating culture and those who concentrate on technology.

This horizontal divide holds practical significance because cultural and technical workers are increasingly part of the same workplace and the same work process. This raises the obvious question of whether they can work together in a harmonious way. More importantly for our purposes, technical and cultural knowledge workers are often included in the same trade unions. Unions like the CWA and the CEP have brought together very different types of knowledge workers, not just those who are higher or lower in a vertical status hierarchy but also those who are on one side or the other of the divide between cultural and technical workers. What does this mean for the prospects of building labor solidarity? Can this succeed within a single union, within a single country, across one border, or across one or more oceans? Considering the vertical and horizontal divisions that mark the meaning of knowledge work, it is easy to appreciate the challenge facing those who would unite them.

In addition to debates about the meaning of the term "knowledge workers," a range of possibilities exists in different interpretations of the term "information society," and particularly its relationship to capitalism.

In the narrowest use of the term, "information society" describes the growth of a particular type of economic activity in capitalist societies: the simple acknowledgment that the production, distribution, and exchange of information have come to occupy relatively more economic activity than either agriculture or manufacturing. In this view, the fundamental nature and dynamics of capitalism—the production of exchange value from use value, and the organization of the economy to maximize surplus value—remain central to the global political economy and, for the most part, unchanged. If anything, the capitalist system of production is now arguably even more consequential than it used to be because the end of Soviet communism has practically eliminated the major challenge to capitalism and because the growing interdependence of the world's economy means that fewer and fewer people can escape the impacts of capitalism. Communication and information technology contribute to capitalism by advancing the process of commodification. The technology deepens, extends, and refines the ability to measure and monitor economic activity, and to package and repackage marketable products. In essence it expands the ability to control time, space, and human activity. Whether the commodity is a material good, an information or communication product, human labor, or consumers and audiences, the development of communication and information technology greatly increases opportunities to turn the atoms and digits of things, knowledge, and people into marketable products. But while the technology greatly expands opportunities to create use value, it does not, according to this perspective, call into question the fundamental nature of capitalism.

Borrowing a phrase from Dan Schiller (1999), the economic system of the information society can be characterized as "digital capitalism." In short, the information society is a capitalist society, first and foremost.

Another view holds that the terms "information age" and "information society" represent a distinctly new form of capitalism, one that simultaneously strengthens some of the fundamental tendencies of the system by expanding markets, but at the same time challenges them. In this perspective, the information society is not simply a capitalist society in which relatively more economic activity is bound up with knowledge and information. Rather, it is a society whose activity raises significant questions for the viability of capitalism because it opens up new forms of production, both within and outside the system. Manuel Castells provides arguably the best current example of work that straddles the fence between a focus on capitalism and on a fundamental break with it.

The fact that Castells (2000, 2003) switches back and forth between calling it "informational capitalism" and a "network society" reveals a genuine concern that, while elements of traditional capitalism remain, we need to be open to the view that the world is experiencing the creation of a fundamentally new type of society. The importance of making the most profitable use of the means of production, including labor, continues, and the social relations of production, if increasingly organized around communication and information, remain distinctly capitalist. However, the enormous and accelerating capacity to create communication and information networks challenges capitalism's capacity to manage and contain them. The volume of information and communication that falls out of the orbit of value extraction—from the simple act of downloading material free of charge, to sending a video message to the world on YouTube, to carrying out criminal activities like money laundering under the cover of cyberspace—threatens the singular dominance of the capitalist mode of production. Open source and hacker networks challenge property rights in the digital world. Criminals and terrorists use digital technology to both hide and extend their operations. Labor unions use networks to coordinate strikes, and to build national or international support for their cause. In essence, the network begins to replace the commodity as the central axis of social development. According to this perspective, the result of all this often-conflicting activity is that we live in a changed world, marked by the tension between information that wants to be free and a capitalism that wants to use it for the singular purpose of creating surplus value.

Castells openly moves between both views, and his work embodies the tensions in the scholarly community over the significance of an information society. Others take a more explicit position that the information society—from its networks that blur the lines between the real and the virtual to its labor that blurs the distinction between material and immaterial—is transforming capitalism into something fundamentally different. (See, for example, Dowling, Nunes, and Trott 2007; Hardt and Negri 2000, 2004; Terranova 2004.) Incapable of containing the process of value creation within the confines of capitalist processes and institutions, capitalism morphs into an information society whose rules and relationships are increasingly up for grabs. According to this view, a society built around information networks erases the once-clear demarcations between work and home, labor and leisure, economic value and social value, and in so doing begins to change into a new kind of society. This opens the way to oppositional and democratic social movements united in their commitment to expand the terrain, both physical and digital, that is not governed by the commodity, the market, and the pursuit of profit. Most of those who maintain this view readily admit that this is certainly not a utopian world; the reliance on intensive surveillance alone challenges that claim. But the growth in the multiplicity and intensity of challenges to traditional capitalism, according to this view, raises fundamental questions that lead some to find answers in a new type of society, an information society.

Thinking about Knowledge Work

Discussion and debate about the relationships between the technologies of communication and information, work, and social development are not new. An extended history of the subject in the West would have to include the work of Saint-Simon (1952), who speculated on the transformation from manual to mental labor in the eighteenth century. It would also include the work of Marx (1973), who referred to the formation of a "general intellect" in the course of capitalist development. Charles Babbage (1832), whose nineteenth-century "difference engine" was one of the first serious designs for a computer, would also have to be included in the conversation. These and a range of other scientists, social scientists, and philosophers have had a substantial impact on how we think about the relationships between work, technology, and society.

For the purposes of this book, however, it is most useful to focus on the debates that accelerated after the Second World War, when scholars began to notice growth in the number of jobs outside the manufacturing sector. In the early post-war years, the academic emphasis was on developing measures to track the growth of the information sector as an economic force. In this regard, Jean Fourastié (1954) provided the first sustained analysis of the division between agriculture, manufacturing, and the expanding services sector. Fritz Machlup (1962) was among the leaders in charting the expansion of the data and information components of the economy. His work received a great deal of attention partly because it defined the knowledge terrain very broadly, mapping both that which is known and the different types of knowers. For the former, he distinguished between several different types of knowledge, including the intellectual or that which results from scientific and humanistic training. He also identified a category of practical knowledge required of those charged with carrying out specific jobs, whether at work, in the home, or in the political sphere. Interestingly, Machlup was among the earliest to reflect on the significance of a third type of knowledge. He called it "small-talk" or "pastime" knowledge, and it would later come to be called the tacit knowledge we take for granted as common sense. He also found places for spiritual knowledge and for "unwanted" or accidentally acquired knowledge. His map of the knowledge terrain was important because it laid the groundwork for paying increasingly serious attention to the role of knowledge and information in society. But Machlup's map of the knowledge landscape was of less interest to him than his examination of the relationship between knowledge and labor.

Machlup's typology of what he called knowledge producers offers a very useful way to think about the variety of people who work in the knowledge trade. We describe it in some detail for that reason, but also because it demonstrates that Machlup's work is worth more than a quick note in a capsule history of information theory. It stands up well as one among other substantial ways of thinking about knowledge labor.

Machlup described the range of knowledge producers this way:

1. Original creators. These are people who draw on a rich store of data but add so much to it in terms of creativity that there is only a weak and indirect connection between what is received from others and what is communicated. Original creators include scientists, artists, and other thinkers who redefine generally accepted ways of seeing things and construct the paradigms that inform the work of other knowledge workers. Original creators provide the core of the creative class of knowledge workers. Most would be included in Florida's designation of a supercreative core.
2. Analyzers. These are people who add judgment and intuition, in addition to a set of established procedures, to the accepted body of knowledge. Journalists, university lecturers, and applied scientists, such as software engineers, would fall into this category today.
3. Interpreters or translators. These are knowledge workers who change the form and content of an original message, but do so in order to create effects that are equivalent to those intended by the original message. Teachers who are not involved in research, language translators, newsreaders, and camera operators are among those who fit this category.
4. Processors. This category includes people who change the form and content of the information store but only by applying routine procedures that

subject received information to specified operations, such as combinations and computations which lead to definite results independent of the processor's own tastes and values. Accountants, news wire service editors, librarians in the circulation and cataloguing departments, and certain lower-level health professionals (those who, for example, read routine scans), provide examples of processors.

5. Transformers. Like the others, these workers also change the form of a received message, but they are not supposed to change the content. Machlup used the example of the stenographers who were more prominent at the time of his writing than they are today. Call-center workers are a contemporary example of transformers because they take a specific written script and are expected to sell a product or offer a service based on it. Individual workers might be tempted to interpret and take other liberties with the message—and they may even occasionally give in to that temptation—but they are made well aware that "stick to the script" is a cardinal rule in the call-center industry.
6. Transporters. It is evidence of Machlup's breadth and perceptiveness that he included those whose job it is to deliver the messages they receive in their precise form as knowledge workers. Messengers, postal workers, and others who take content from one place and deliver it to another make up this group.

One of the challenges facing knowledge workers today is whether concrete connections can be made between the experiences of transporters and those at the higher end of the knowledge hierarchy. For example, can higher-level knowledge workers like writers and programmers draw lessons from the successful 1997 Teamsters' strike against United Parcel Service? (See Kumar 2007a, 2007b.) Are these various groups similar enough to work together successfully under the banner of the same trade union or worker association? This is not a matter of idle speculation; as chapter 6 describes, the association of computer professionals known as WashTech succeeded, with the help of the CWA, in organizing workers at the Cingular wireless company.

Machlup made a significant contribution to the understanding of knowledge and of knowledge labor, but he admitted that he could not carry out the job of counting knowledge workers. "What I have failed to do . . . is to attempt a statistical separation of knowledge-producing activity by these different levels [transporters, transformers, processors, interpreters, analyzers of messages, and original creators]" (1962, vi). That awaited the work of Marc Uri Porat.

Porat (1977) documented the shift from an economy based on the primary (agriculture) and secondary (manufacturing) sectors to one rooted in the tertiary and quaternary sectors, where services and information predominate. In addition to building a statistical model based on an analysis of four hundred occupational types reported by the U.S. Census and the Bureau of Labor Statistics,

Porat sought to overcome another weakness in Machlup's work. Machlup failed to document those difficult-to-account-for information activities that are contained within the in-house elements of companies whose primary product is not in the knowledge sector. So, in addition to the primary information sector which included the mass media, education, and computer manufacturing industry where the value of the information produced can be easily identified, Porat identified a secondary sector comprised of the information units of noninformation companies, such as the research and development arm of a pharmaceutical company. When he put these together, Porat concluded that information activities comprised fully 46 percent of U.S. gross domestic product (GDP).

In spite of the improvement over Machlup's assessment, Porat left himself open to at least two types of criticisms. In the effort to account for as much of the information content in the economy as possible, Porat erred on the side of inclusiveness. By extending his search into the knowledge-intensive arms of companies outside the primary information industries, Porat raised the question of where the search ends. Isn't there an information component to every job? Doesn't even the automobile assembly worker have to draw on a store of information to make decisions about manual actions? What about the cable repair technician faced with a faulty signal? The second concern has to do with the question of power. Does it make sense to include the equivalent of a chief information officer and a file clerk who works in a research division in the same corps of knowledge workers? Surely it matters more that they are at opposite ends of a corporate ladder than it does that they both contribute to the information sector's share of GDP.

The point of these criticisms is not that Porat's estimate is insignificant, but that it is overemphasized. Also, there are more important goals to achieve in this field. Rather than argue over how much information labor one might squeeze out of the occupational structure to understand GDP composition, it is more important to understand the nature of information labor and the convergences and conflicts that result from its development. Here, Porat made a contribution similar to Machlup's by providing a typology of information work. Porat recognized that making distinctions, while essential, was a "risky proposition" (1977, 5). He offered a three-part scheme that begins with a primary category of workers whose job is to produce and sell information. This group includes scientists, inventors, teachers, authors, journalists, and librarians. The secondary knowledge-worker category encompasses those who gather and disseminate the information that primary producers distribute and sell. For Porat, this would include managers, secretaries, clerks, lawyers, and stockbrokers. Finally, Porat offered a category of information technology operators, such as computer operators, cable installers, and television repair workers, who make the system work.

Machlup and Porat made important contributions, especially for understanding the expansion of knowledge labor and its structure. But neither addressed the political, social, and cultural implications of this transformation in anything approaching the theoretical sophistication of Daniel Bell (1973). Bell was not alone, as the work of Touraine (1969/1971) and Gouldner (1979) attests. But the Harvard sociologist provided the most thorough analysis for his time, and his work served as a lightning rod for criticism and further development of the field.

According to Bell, the growth in data and information work and the changes this brought to the occupational structure in America were significant developments. Drawing on the work of Foote and Hatt (1953), who provided one of the first economic mappings of the information society, Bell (1973, 35) described the "crucial importance of the quinary or intellectual sector." But Bell argued that something far more important than a simple shift in the occupational structure was taking place, and this was nothing short of a transformation in the nature of capitalist society. Bell began his assessment of the movement toward a postindustrial society by identifying five central characteristics and using them to understand what was happening to capitalism. At the base is the creation of a service economy, one in which the majority of labor is no longer engaged in agriculture or manufacturing but in the service occupations that encompass trade, finance, transportation, health, recreation, research, education, and government. Among these, Bell saw the growth in occupations covering health, education, research, and government as most decisive because "this is the category that represents the expansion of the new intelligentsia" (1973, 15). The second key development Bell identified is a changing occupational distribution that gives preeminence to a new professional and technical class. For Bell (1973, 17–18), occupation is the most important determinant of social-class stratification in society, and the most "startling" change he observed is the growth in the category of college-educated people, primarily in the following categories:

1. Science and engineering: natural scientists, social scientists, engineers.
2. Technicians.
3. Medical health: physicians and surgeons, nurses, dentists, pharmacists, psychologists.
4. Teachers: university, secondary, elementary.
5. General: accountants, clergy members, editors and reporters, lawyers and judges, arts and entertainment, architects, librarians, social workers.

In addition to a significant new economic sector and a changing occupational and class structure, Bell argued for a new "axial principle," or organizing force, in capitalism: the primacy of theoretical knowledge. Not only had knowledge become more important, but the specific type of knowledge that is central to capitalist society had changed. Theory was taking precedence over empirical knowledge. Specifically, social management, growth, and control would come to be based on the codification of knowledge into abstract systems of symbols that are used to understand diverse kinds of experience. Equally important was the growth in a future orientation, seen most specifically in the impetus to engage in technology planning. For Bell, this presaged a novel phase in human history: the conscious planning of technological change that would ameliorate the uncertainty that had always been a worrisome characteristic of economic life. Finally, just as important as the growth in material technology was the advance in intellectual technology. By this he meant the development of the tools, or what we would today call the software, to manage the complexity of advanced industrial societies. These developments would undermine the social relations of industrial society to such an extent that capitalism itself would be transformed.

Capitalism had been governed for two centuries by industrialists and their financiers, who comprised the capitalist class. But with the rise of a society dependent on intellectual technology, and particularly on the production and distribution of theoretical information, Bell maintained that a new class of leaders was rising to prominence. This was a genuine knowledge class of well-trained scientific and technical workers who ultimately would assume leadership of a postindustrial capitalism. As inherited wealth and power shrank in significance, a genuine meritocracy would eventually come to rule. Such a society would not necessarily be more democratic, but it would certainly take on a different character as power shifted from its traditional base in family inheritance to intelligence based on an education rich in technical and scientific knowledge. The growing ranks of knowledge workers would literally power and manage this new postindustrial economy, leading to steady economic growth and the decline of historic ideologies. According to Bell, political battles over public policy would diminish as technical algorithms and knowledge-based measures would come to govern.

There would no doubt be tensions in such a society, but they would be technical in nature rather than ideological. In Bell's postindustrial world, the only potential for serious division lay outside the economic and political spheres, residing in—as Bell would argue in his next, far darker book, *The Cultural Contradictions of Capitalism* (1976)—the cultural sphere. The danger would come from a culture sinking deeper into consumer hedonism and captivated by irrational beliefs. Together, these two seeming opposites—the growth of materialism and of a counterculture resistant to the disciplines of a capitalist workplace—could threaten the foundation of postindustrialism because they challenged the delayed gratification and support for technical rationality that were required to maintain it.

It did not take long for others to conclude that, cultural issues aside, postindustrialism was not an inherently progressive force. Two examples, published within a year of Bell's *The Coming of Post-industrial Society*, suggest the scope of this critique. For Herbert Schiller (1973), postindustrialism meant the rise of transnational media and communication businesses that were pumping out support for American values, including U.S. military and imperial ambitions, and eliminating alternatives through increasingly concentrated market power. Schiller agreed with other scholars that the occupational structure was changing, but only in the sense that corporate power was shifting to companies in the media and information sectors. The names of leading corporations and the job titles of their employees would change, but the class structure remained about the same. Next, Harry Braverman (1974) argued that the fundamentals of capitalist control over work—including the effort to take apart the labor process as a process controlled by workers and then reconstitute it as a process controlled by managers, and the drive to use technology to deskill and ultimately supplant living workers with "dead labor" through the development of machines that embody their skills—remained unchanged. He warned that for the vast majority of workers in the service, retail, and knowledge professions, their labor would become about as regimented, and ultimately deskilled, as the labor of those in assembly line manufacturing. Indeed, given the immateriality of knowledge work, it would be easier than in the industrial era to separate conception from execution, and to concentrate the power of conception, including design and management, in a dominant class.

There has been widespread debate over the postindustrial society ever since Bell, Braverman, and Schiller addressed these issues in the early 1970s. Important contributions have been made to the literature using a wide range of methods, from the historical (Blok and Downey 2003) to the ethnographic (Barley and Kunda 2004; Marschall 2002). But there is some agreement in key areas. There is consensus, for example, that a shift from manufacturing to knowledge work has occurred in the developed countries of the West, and is beginning to occur in the other parts of the world. People agree that a considerable amount of knowledge was, and still is, required in much of manufacturing, as well as in agricultural work. In fact, the information component of the work in these sectors has arguably grown substantially in the most developed nations. But today, an increasing amount of work is taken up with the production and distribution of information, communication, and knowledge. Furthermore, there is agreement that a dynamic process of deskilling, upskilling, and reskilling is taking place in the occupational hierarchy. At different times and in different sectors, one or another of these processes predominates, although the labor process, most concur, cannot be reduced to the singularity of one standardized set of activities (Brint 2001; Huws 2006; Powell and Snellman 2004).

Zuboff's (1988) research in a variety of work sites demonstrated that companies were capable of succeeding by increasing the skills of their employees, a process she referred to as "informating" or strengthening workers' informational—and, at times, even their conceptual—capabilities. Nevertheless, there is also agreement that companies have benefited from reducing the skill component of jobs and from eliminating jobs entirely and replacing them with automated systems. This applies especially to jobs traditionally filled by women who make up what Huws (2003) calls the new "cybertariat." (See also Tracy and Hayashi 2007.) Scholars also concur that there is evidence of a polarization in the workforce as middle-level positions are reduced or eliminated, and as credential barriers have been erected to separate higher- and lower-status occupations. One consequence of this is the reinforcement of racial and sexual barriers in the workplace (Burris 1998).

Where deskilling or job elimination is not possible, companies have accomplished the same cost-cutting objective by moving jobs, either to low-wage (and often nonunionized) areas within a country or by sending them abroad. Since knowledge work typically does not require moving material things over long distances, it is especially vulnerable to the practice that has become known as offshore outsourcing or, simply, outsourcing. This process is an extension of the business-led, neo-liberal agenda that has transformed the business-labor social contract of the 1950s and 1960s. That contract promised workers secure jobs at a living wage with a package of benefits in return for ceding complete authority over the economy to business. Today's corporate organizational and operational structure sacrifices jobs on the twin altars of productivity and profit. Outsourcing is not without its antinomies, as chapter 7 describes, and there is considerable debate in the literature on their significance, but there is no dispute over its increasing significance for knowledge work. The immateriality of knowledge work reduces transportation costs substantially, leaving the primary challenge, admittedly often a significant one, of how to communicate successfully in a world of cultural difference.

Not all contemporary theorists see immateriality as a negative for labor. Scholars like Hardt and Negri (2000), Terranova (2004), Dyer-Witheford (1999), and Lazzarato (1997) agree that knowledge work or immaterial labor—in other words, work that produces a service, a cultural product, knowledge, or communication—is increasingly central to the global economy. Moreover, and this is one of the reasons the focus of these "autonomist" thinkers is on immaterial rather than just knowledge labor, much of the laboring draws on affective skills, not just informational skills. Workers have to learn how to emote as well as think; they must produce feeling and empathy as much as they do knowledge and information. The totalizing nature of immaterial labor opens new possibilities for new levels of control over labor, as managers can now routinely make affective as well as informational demands on workers. But as these writers point out, this also means that cooperation is essential to the contemporary labor process because affective labor demands human contact and interaction. As a result, workers gain the possibility of valorizing themselves collectively in opposition to capital, as well as through new forms of labor that operate outside and therefore challenge the system of surplus value creation. As Hardt and Negri (2000, 294) put it, "the cooperative aspect of immaterial labor is not imposed or organized from the outside, as it was in previous forms of labor, but rather, cooperation is completely immanent to the laboring activity itself." With productivity, wealth, and the creation of social surpluses completely dependent on multiple forms of cooperation and interaction through linguistic, communicational, and affective networks, the information society implicitly provides workers with skills that can be used to oppose the logic of capitalism, not simply to advance it.

The Concept of Convergence

In addition to examining the range of ways to think about knowledge workers and the nature of the information society in which they labor, it is important to consider the concept of convergence, a term that came to prominence in the 1990s. As one of the central concepts used to describe developments across the media, telecommunications, and information sectors of the communications industry, convergence is also central to understanding knowledge work and knowledge-worker organizations. Pavlik and Dennis (1993, 2) define convergence this way: "the integration of or interface between and among different media systems and organizations, made possible by the development of new technologies." More generally, the term refers to the integration of technologies, arenas, and institutions in the industry (Babe 1996; McKercher 2002; Winseck 1998).

Technological convergence typically means the integration of the devices that these industries use, as well as the information they process, distribute, and exchange over and through these devices. Bringing together computers and telecommunications and built on a universal language of digital code, the Internet is an iconic example of technological convergence. Technological convergence is linked to, and partly responsible for, the convergence of once-separate industries into a common arena providing electronic information and communication services. At the same time, corporate convergence promotes, and is promoted by, technological convergence.

Technological differences once erected thick walls between print media, electronic media, telecommunications, and information services. The barriers between various media forms were sufficiently well defined that governments around the world created separate legal regimes to deal with each of them, often regulating broadcast media in different ways than print media, for example. The barriers extended to the labor process and to trade union structures in those industries. Producing a newspaper required a diverse set of workers whose skills were specific to the various stages of the production process, from setting type to making engravings of photographs to manufacturing printing plates to operating the press. These workers were represented by an equally diverse set of trade unions, who divided jurisdiction along the lines of the production chain. Similarly, producing a broadcast program required workers with specific and highly developed sets of skills, ranging from collection of sound to processing of film to melding audio and visual elements into a seamless package. Some skills were common to both media forms, such as collecting interview material or editing stories, but required mastery of vastly different sets of tools. In recent years, however, all those walls have been breaking down, eliminating many of the distinctive features that kept these industries divided and creating one large electronic information and communication services arena.

Convergence has enabled the interconnection of technologies to create new systems of hardware and new levels of service, such as wireless networking. These forms of hardware convergence are greatly advanced with the development of a common digital language that does not distinguish between audio, video, or data transmission, reducing all communication to one language that provides a manifold increase in the quantity and quality of electronic communication. Digitization has the technological advantage of providing enormous gains in transmission speed and flexibility over earlier forms of electronic communication, which were largely reliant on analog techniques (Longstaff 2002). But it takes place in the context of, and greatly expands, the process of commodification, that is, the transformation of what amounts to a resource into a marketable product or service. The expansion of the commodity form provides the context for who leads the process of digitization and how it is applied. In this way, the commodification of information and entertainment not only expands the market for communication products but also deepens the commodification of the labor involved in the production, distribution, and exchange of communication. Finally, this process furthers the growth of markets in the audiences that receive and make use of electronic communication (Mosco 1996).

Digitization serves to advance the overall process of commodification worldwide and its specific application to communication. It does so by extending the range of opportunities to measure and monitor, and to package and repackage, information and entertainment. A similar extension of commodification applies to the labor of communication. Beginning in the 1970s, the replacement of mechanical with electronic systems transformed the labor process in the print and broadcasting industries. Technological change took on a routinized and unrelenting character as employers applied digital technology to almost every aspect of production. In newspapers, digital innovations hollowed out the production process, collapsing previously separate stages into each other and eliminating previously distinct crafts. Digital systems continue to allow companies to expand this process, enabling centralized customer service facilities, for example, or regional printing plants for newspapers owned by the same corporation. Newspaper journalists increasingly serve in the combined roles of editor and page producer. They do not simply report on a story; they put it into a form for transmission to the printed page and, increasingly, to the electronic page. Similar patterns are found in broadcasting. TV news crews used to rely on separate craft workers to collect images and sound, and on specialized technicians to help reporters and producers transform the material they collected into stories. Today's broadcast journalists have become digital videographers, shooting and editing their own material for delivery over television or computer networks. Meanwhile, newspaper and broadcast workers find themselves using the same tools to edit sound, pictures, or text. The film industry is experimenting with delivering digital copies of movies to theaters in multiple locations over communication satellites, thereby eliminating the distribution of celluloid copies for exhibition by projectionists. Software is sold to customers well before it has been debugged on the understanding that customers themselves will report errors and download and install patches and other corrections, as well as figure out how to work around problems. The ability to eliminate labor, combine it to perform multiple tasks, or shift work to unpaid consumers further expands corporate revenue opportunities (Balka 2002; Liu 2004; McKercher and Mosco 2007).

Companies are taking advantage of technological convergence by creating corporate or institutional convergence. This is embodied in the scope of merger and acquisition activity that, while not limited to the communication industry, is arguably most prominent within it (Klinenberg 2007; Mosco 2004; Nichols and McChesney 2005; Schiller 1999). In the United States, the media landscape is dominated by a handful of corporations, including a reconstituted AT&T; Time Warner; GE, the parent of NBC; Disney, which controls ABC; News Corporation, which owns Fox; and technology leaders Microsoft and Google. These companies have converged vertically, controlling the various pieces of the news and entertainment products they make from the beginning to the end of the production process, and horizontally, by extending their control across practically all sectors of the industry. They have also sought to extract maximum value from individual properties by converging production, marketing, and customer service facilities into centralized hubs.

This development is not limited to the United States. Corporate convergence is further along in Canada, for example, where four firms—Bell Canada Enterprises (BCE), Rogers, Quebecor, and Canwest—are dominant in journalism, broadcasting, cable, and telecommunications. Other companies, including Telus and Shaw Communications Inc., are also expanding, in hopes of challenging the giants. But concentration is growing and there is little question that convergence is bringing together communication firms that want to take advantage of opportunities to integrate products and services, to cross-produce, cross-promote, and cross-market in previously separate spheres like entertainment and news.

Corporate convergence does not, in and of itself, guarantee success. Nor does it magically produce the synergies that companies anticipate, such as integrating the cultures of the print newsroom and the broadcast station. It sometimes results in content that cannot attract audiences. These facts help to explain the difficulties experienced by convergent media firms like Time Warner, which continues to struggle with how to integrate its AOL division, or BCE, which seems to be embracing convergence and divergence with similar levels of enthusiasm. Moreover, digitization itself is not a flawless process. Technical problems may slow its development, and solving one technical problem can sometimes create other problems. Another stumbling block affecting the process of technological and institutional convergence is government regulation. In the late nineteenth and early twentieth centuries, government regulators helped erect and reinforce the walls that separated media forms from one another. They did so for a number of reasons, including concern over monopolies and a need to bring order to the allocation of the broadcast spectrum. The current wave of technological and institutional convergence raises fundamental problems for regulatory policies that are based on discrete industries and discrete technologies. And while the tendency of government regulators in Canada and the United States in recent years has been to move toward deregulation and the easing of constraints on cross-media ownership, this has not happened smoothly, or overnight. But these may be short-term problems, resulting in cyclical declines over the course of a secular trend, rather than evidence that convergence has failed. Large units enable businesses to better control their environments, limiting competitive pressures even as they benefit by developing internal market competition among divisions.

Convergence is not just a technological, political, and organizational process. It is also a myth or a story about how computer communication is revolutionizing technology, politics, and society. As such it is part of a sublime vision that, in its strongest form, imagines that technology can create the conditions for the end of history, the end of geography, and the end of politics (Mosco 2004). Convergence is therefore more than just a term to describe an ostensible change in technology and organization. It is part of a utopian discourse that aims to lead us from the coarse materiality of, in Nicholas Negroponte's terms, "the world of atoms" to help us "learn to be digital" (Negroponte 1996). This affirmative vision is used to rationalize deepening social inequalities like the digital divide between the information-rich and the information poor, tightened surveillance practices, and the growing control exerted by a handful of companies over the production and distribution of communication and information. To say that convergence is a myth is not to imply that it is false. Rather, myths take a basic empirical reality and enlarge it by attributing transformative social and cultural consequences that are not currently justified by empirical evidence.

As both a political and a cultural process, convergence creates considerable pessimism among those who support democratic communication, diversity in the form and content of information and entertainment, and universal and equitable access to media (Artz and Kamalipour 2003; Herman and Chomsky 2002; Winter 2005). But workers are responding to convergence with a form of their own that holds promise for the labor movement, especially among knowledge workers.

Labor Convergence

Chapter 5 takes up case studies in labor convergence, but this is a useful point at which to examine the term analytically. As the previous section shows, convergence does not simply mean an increase in size. Rather, the word implies that integration or interface involves a transformation of some sort, and an opening up of opportunities as a result of a new configuration. This applies to labor convergence as well. In addition, because the term is connected to developments in the knowledge-based economy and society, we feel it is particularly applicable to the unions that represent communications workers.

There is considerable research on the value of convergence among trade unions, including in the communication and information industries (Batstone 1984; Katz 1997; McKercher 2002; Stone 2004). To a degree, converging unions see these actions as defensive, or as ways of protecting their members from the growing power of companies that employ them and from governments that have, more often than not, taken a probusiness position. But significantly, they also see union convergence as an attempt to take advantage of synergies brought about by growing convergence in their work (Bahr 1998). Since communication workers unions represent people who are increasingly involved in producing for a converging arena of electronic information services or knowledge, these unions see improved opportunities for organizing, for bargaining, and for successful political intervention. Labor convergence, therefore, covers a range of activity, including mergers between previously separate trade unions, attempts to use bases in one industry or trade to organize in others, efforts to consolidate bargaining units in companies where multiple unions are the norm, and the possibility of building bridges to organized and unorganized workers at home and abroad.

As with the corporations that employ them, there is a mythic aspect to the notion of labor union convergence. Recalling the heyday of labor organizing by the Knights of Labor in the late nineteenth century and the explosion of industrial unionism in the 1930s, today's unions are reviving the historic vision of one big union, which imagined bringing together workers across all of the great divisions that traditionally separated laborers and constrained opportunities for resistance and the pursuit of a labor-centered agenda. The unions promoting the ideal, like the CWA and the global federation Union Network International, modify it to fit the vision of an information society. But whether calling for workers of the world to unite or concentrating only on uniting knowledge workers, the goal of transcending current struggles and setbacks with a mythic vision of a new day for labor continues to resonate. In essence, converging technologies, converging jobs, converging companies, and the belief that one big union is better have led workers to come together across the knowledge industry (McKercher 2002). This has strengthened the position of labor as a source of resistance to the erosion of public services, including in the communication arena, and has made it all the more important for companies to fight against convergent unions.

This does not mean that trade union convergence is the answer to all of labor's ills. In certain cases, breaking away from a converged union or labor federation can benefit at least some members. For example, research has demonstrated that unions which emphasize organizing and adding new members over providing services to existing members operate more effectively when they remain small and independent from other unions (Fiorito, Jarley, and Delaney 1995). Indeed, the argument for the value of smaller and perhaps more flexible, agile, and generally less bureaucratic structures underlies the formation of the breakaway Change to Win coalition in the United States, which pulled a number of unions out of the AFL-CIO in 2005. In addition, some Canadian trade unionists have come to believe that national trade unions serve their needs better than the so-called international unions based in the United States, and have broken away from the internationals to form their own unions. Several factors account for this, including a desire for autonomy, concern about the increasing divergence of labor relations regimes in the two countries, and differences over political action strategies (Dawes 1987). This trend is very apparent in the communication and information sector. The "communications" part of Canada's CEP union, for example, was originally the Canadian arm of the CWA. CEP calls itself Canada's media union, but the bulk of its media workers come from locals that have separated from U.S.-based unions, including the Newspaper Guild, the International Typographical Union, and the Graphic Communications International Union.

Many employers prefer smaller bargaining units that have not joined other unions in convergence on the grounds that small, independent unions typically lack the power and resources of larger ones. This enables companies to benefit from a weak employee organization even as they implement what Keefe and Batt describe as centralized decision making that seeks to take advantage of "scale economies and cutting costs through consolidations, new applications of technology, reengineering, and downsizing" (1997, 67). These authors conclude that such pressures have overtaken almost every effort to apply genuine, employee-led strategies that would incorporate locally designed job enhancement and employment security measures across the U.S. telecommunications industry. Moreover, in multiple-unit settings, where unions have done little to organize and bargain together, employers can play one unit off against another, isolate units, and otherwise whipsaw unions into concessions.

Nevertheless, there are times when this strategy can backfire for business because it increases the sheer amount of bargaining as companies have to negotiate with several different units rather than with one. In addition, unions gain the opportunity to whipsaw management: if one unit wins a concession, the others may seek it too. Moreover, there is no guarantee that multiple units create more stable employer-employee relationships because bargaining with multiple units can heighten tensions between groups of employees, who may feel that their unit or department faces more stringent bargaining pressures than other units or departments. As a result, the overall stability of employer-employee relationships can be threatened. Finally, bargaining divided among multiple units makes it more difficult for businesses to make the best use of their human resources because it creates chains of opportunity and mobility within units rather than across units. As a result, some employers have sought bargaining-unit consolidation. For example, in the communication sector, these include Canada's national public broadcaster, the CBC, which has been consolidating bargaining units in stages over the last fifteen years, and the Pacific Newspaper Group, which publishes the leading daily newspapers in Vancouver and applied successfully for consolidation of seven bargaining units into one in the 1990s.

Concluding his survey of the bargaining structure literature, Anderson (1989, 231) acknowledges that convergence is good for the general system of industrial relations and not just for labor. He finds that it benefits the overall economy because it reduces industrial conflict, strengthens professionalism, and eliminates settlements that are out of line in favor of either party. In some ways this approach resurrects the labor-capital social contract that predates the emergence and acceleration of neo-liberal reforms. It also harkens back to the ideas embodied in the New Deal legislation of the 1930s, which reacted, in part, to the tumultuous labor relations practices that dotted the industrial landscape of the nineteenth and early twentieth centuries. Commenting on efforts to shift from centralized bargaining among converged units at British Telecom (BT) in the United Kingdom, Katz (1997) concludes that companies have come to appreciate the value of the old system. He notes that for BT, early efforts at divided bargaining did not work for management because the company ran up against the need for standardized services and maintenance of a coordinated telephone service. BT management came to appreciate, as did other companies, the stability and predictability that pattern bargaining and a high degree of standardization in collective bargaining agreements can bring to the workplace. Specifically, convergence, which manifested itself in centralized and standardized union structures, satisfied the need for consistency and coordination that arises from the nature of the core telecommunications network (Katz 1997, 21). Keefe and Boroff (1994, 347–48) came to a similar conclusion in their analysis of collective bargaining in the U.S. telecommunications industry one decade after the AT&T divestiture.

According to Anderson, divided bargaining does not help labor because it leads to the development of a status hierarchy within the company that benefits some workers at the expense of others. It also creates opportunities for whipsawing or bargaining that plays one unit of labor off against another, thereby fomenting tension and conflict within and between different groups of workers. Drawing on research from four different sources, Anderson (1989, 220) concluded that divided structures allow management more opportunities to defeat unions through relocation, technological change, and other cost-control measures. It is difficult to understand how an approach that is set up to replace and avoid unions, among other things, can benefit labor and labor relations.

Two leading industrial relations experts, Anthony Ferner and Michael Terry (1997), concluded their research on British Telecom with the finding that BT's experience with divided bargaining amounted to an attempt by the company to pursue "divide and rule." This met with short-term success but created enough turmoil for overall industrial relations within the company that BT returned to a more centralized pattern of bargaining. Notwithstanding certain short-term gains for management, communications firms are learning the lessons of earlier industries that attempted to divide bargaining among smaller and smaller units. The gain from dividing and weakening labor does not necessarily outweigh the negative consequences for the overall integrity of industrial relations. These consequences were pointed out in 1968 by a United Kingdom Royal Commission (the Donovan Commission), which reported on major defects in what was then the fragmented British industrial relations system:

> the tendency of extreme decentralization and self-government to degenerate into indecision and anarchy; the propensity to breed inefficiency; and the reluctance to change. All these characteristics become more damaging as they develop, as the rate of technical progress increases and as the need for economic growth becomes more urgent. (United Kingdom 1968)

Keefe and Boroff conclude that as a result of this recognition, even firms in turbulent and competitive environments, such as the communications industry, are recognizing the values of long-term stability and labor harmony—things that convergence and centralized bargaining help to bring about.

Experts on trade union structure and on the relative merits of convergence versus divided bargaining units also recognize a gain for consumers in relatively centralized structures that are the products of convergence. According to Batstone, converged unions take on the character of "encompassing organizations." As a result, depending on strategies that unions and other social actors pursue, union leaders in converged organizations have to take greater account of the ways in which union victories can impose costs upon their members when they take on the role of consumer (Batstone 1984, 330; see also Anderson 1989). There are labor relations benefits for trade unions that pursue a strategy of convergence. That there are benefits for labor harmony and for consumers as well helps to explain why farsighted government labor boards also support bargaining-unit consolidations.

Thinking Critically about the Importance of Knowledge Labor

As knowledge workers themselves, the authors are acutely aware of the potential to both exaggerate and to underemphasize the importance of knowledge labor. Scholars may risk overstating its importance because knowledge workers form a group that is closely tied to our own professional and political identities. It always feels a bit uncomfortable when intellectuals write about the increasing centrality of their class without much self-conscious reflection on how their views might be skewed by their own position as knowledge workers. Therefore, we feel it is important to explain why knowledge labor takes a central place in a book about worker organizations.

Without doubt, manufacturing labor remains a significant force even in the most developed societies, and manufacturing workers unions, however far they have slipped, remain an important political force. In 2007 the United

Steelworkers (USW) in the United States and two industrial unions in the UK, which were themselves in the process of merging, decided to pursue an international merger encompassing all three of them. This demonstrated that manufacturing unions are seeking to assert themselves in new and creative ways, not just nationally but internationally. In announcing the plan, the unions noted that the resulting organization would be the first trans-Atlantic labor union and, with 3.4 million members, the largest trade union in the world (USW 2007). It would also be a union that is active in a tremendously broad range of industries. The Steelworkers already represented workers in pulp and paper, forestry, rubber, energy, and general industry. The new British superunion, known as Unite—formed by the merger of Amicus and the Transport and General Workers' Union—represented workers in manufacturing, engineering, finance, transport, the motor industry, energy, construction, health care, agriculture, food and drink, chemicals, and pharmaceuticals (Buckley 2007). In addition, there has been talk about extending this embrace to other unions, including IG Metall in Germany and the International Association of Machinists and Aerospace Workers (IAM) from the United States (O. Morgan 2006).

Certainly, a development like this is worthy of attention. So why concentrate on knowledge workers? We do so because knowledge workers take up an increasingly large share of the jobs in the developed world, and their numbers are swelling in the poorer nations as well. As a result, knowledge workers will play an increasingly critical role in the labor movement in North America and abroad. A labor movement established in the struggles of an industrial era will fail if it neglects what is arguably the largest and fastest-growing component of the workforce. Moreover, knowledge work is central to the transformation that is taking place in the international division of labor. A labor movement that simply opposes the exporting of jobs will fail if it neglects to build bridges to workers in developing nations who benefit from offshore outsourcing. Furthermore, an investigation of knowledge work leads to a reconsideration of an old question for a new time: the relationship between manual and mental labor. A labor movement built on blue-collar workers learning how to win struggles against white-collar managers will fail unless it faces up to the challenges of what Andrew Ross (2003) calls a "no-collar" world. Finally, the labor movement is only now beginning to confront a question that will influence its success for a long time: Can it overcome the divisions among knowledge workers, and differences in the nature of work, in skill levels, and in the traditions and cultures of labor-management relations?

At the same time, we must resist underemphasizing the significance of knowledge workers and their importance for the labor movement because we don't want to be accused of utopian thinking. Myths that conjure one big union of knowledge workers have their value, but can get seriously in the way of thinking clearly. One way to address this problem is to define major ideas broadly and to introduce concepts that provide the grounding for a critical edge. This leads us to address not only the question of whether knowledge workers will unite, but what will happen—and what should happen—if or when they do. We conclude this chapter by looking at four such concepts that enable us to enlarge our vision of knowledge workers in society: technology, communication, democracy, and citizenship.

It is tempting to define technology as a thing that shapes or determines occupations and their hierarchy. In this sense, knowledge workers rise up the occupational ladder and, more importantly, take a central role in the social power structure by virtue of their mastery over technology. The tendency to conflate technology with material things arises out of lived experience and a commonsense view of the world, where we encounter technologies as objects: iPods, lawn mowers, television sets, nuclear reactors, microwave ovens, and so on. A less obvious way to think about technology, but one which students of the subject have found to be more useful than the strictly materialist view, is to see technology as a congealed social relationship (Woolgar 1996, 89). This idea holds that technology congeals in material objects the sets of social relations that enter into its production and use. Specifically, technology crystallizes business and investment decisions and the labor of its designers, producers, and ultimately its users, who take what labor and capital have built and bend it, sometimes easily, sometimes with great difficulty, to their specific needs. Moreover, even in its congealed state, technology is dynamic. The specific ways users mold technology feed into new rounds of decisions by designers, investors, and workers, reshaping a technology when necessary to meet a newly developed (or newly created) set of needs. This explains why the mobile phone began as a tool for voice communication but rapidly became a device for communicating by text, taking still or motion pictures, uploading and downloading digital files of just about any sort, surfing the Internet, or passing the time by playing games. Similarly, the decision to mass-produce personal computers was not one that came quickly, even to leaders in the field like Bill Gates. Initially believing that there would be no need to produce more than a handful of computers worldwide, Gates had to change his view when users responded positively to the first, admittedly primitive, personal computers, and when designers came up with new ways to use them. What is especially relevant to this analysis is that by defining technology as social relations embodied in material objects, we are led to broaden our conception of the social space that technologies encompass. It now includes the full range of laborers, from those who design machines to those who solder chips to a motherboard to those who work with the finished machines, and it also includes consumers and investors.

Communication is a term that is often defined as a product of media and information technology. We talk about print, broadcasting, telecommunications, and the Internet as forms of communication. In this sense, communication is conflated with media that help to produce and distribute communication messages or information. Again, based on lived experience and a commonsense understanding of the world, this is understandable, especially when we consider the pervasiveness of newspapers, radios, televisions, telephones, and computers. But particularly when the discussion turns to technology and information, it is more useful to broaden that view, to insist that communication means much more than what a medium produces, presents, and delivers. Our conception of communication follows from a general interest in placing social processes and social relations in the foreground of research (Mosco 1996). It therefore begins with the idea that communication is a social process of exchange whose product is the mark or embodiment of a social relationship. Material technologies can certainly advance communication, as anyone who has kept in close contact with a distant friend by telephone or e-mail will attest. But they can also get in the way of communication, as anyone who has tried to carry on a meaningful conversation with someone glued to the television would likely insist. Regardless of whether technologies facilitate or get in the way of communication, at its core, communication is a social process. Broadly speaking, communication and society are mutually constituted: the exchange of messages that communication embodies is a process that interacts with and helps to constitute the network of relationships and institutions that we typically call society. Another way of understanding this is to see communication and society as dialectically intertwined. This "social" definition is important because it means that the path from communication to democracy, as well as to citizen engagement, runs through society first and foremost. Technology plays a role, but that role is complex because technology can dissipate as well as enhance democracy. Giving people computers in the hopes of expanding democracy, an idea that forms the basis of many progressive proposals for communication policy, often leads to disappointment for the simple reason that more technology does not guarantee quantitatively more or qualitatively better communication. Communication itself is no guarantee of democracy—just think about how mass-communication technologies have served politically repressive regimes—but when we expand the social capacity for communication, when we expand opportunities for social exchange, we are more likely to succeed in encouraging more democratic communication.

Democracy is another complex notion, one that is also often conflated with technology. The more people use the Internet or the cell phone, the story goes, the more likely they are to participate more actively in social life. In addition, the definition of democracy is often narrowed to a particular practice, voting, that is merely one embodiment of democracy. According to this view, societies are considered democratic when their citizens have the right to vote. The same goes for trade unions, which are seen as legitimately representing their members' needs when they hold elections for union officials. Voting is important, but it does not fully encompass the meaning of democracy. We need only think of the number of fraudulent elections we have seen over the years, where the casting of ballots has been used to paper over and perpetuate corrupt and repressive regimes. Therefore, we need to approach democracy, as we approached communication, as socially grounded and profoundly broader than the act of voting. Specifically, this means seeing democracy as the fullest possible public participation in the decisions that affect our lives. This includes the political arena but it also means the fullest possible participation in the economy, in society, and in culture. Certainly, the act of voting is an important social assertion of political democracy, but its ability to bring about democracy is limited when wealth is concentrated in the hands of a few. In other words, concentrated market power diminishes the popular power that voting is supposed to bring about. Similarly, when cultural power is concentrated in the hands of a few, as is the case when a handful of companies own a nation's major newspapers and broadcasting and telecommunication companies, voting is also distinctly limited in what it can accomplish. This means ideological hegemony is perpetuated and defended by the same power brokers who control the most substantial means of communication. For democracy to flourish, society needs to create the conditions for relative equality and civic participation across the political, economic, and cultural spheres. Extremes of wealth and deep divides in power make full public participation in decision making nearly impossible, including participation in worker organizations and trade unions. But even when relative equality applies, societies face a risk of democracy stagnating without the institutions and traditions that make widespread participation possible.

The final term to consider is "citizenship" and, as with the others, our goal is to broaden the commonsense conception of its meaning and give it a distinctive social and cultural grounding. T. H. Marshall (1964) made what has become the standard contribution here and our intention is to build upon it. He noted that citizenship began as a legal notion that conveyed certain rights, including such fundamental ones as habeas corpus, due process under the law, the presumption of innocence, and the right to a trial by a jury of one's peers.

Over time and much social struggle, these legal rights were extended into the political arena, with the right to vote and to participate in the political process. Today, the struggles center on extending these rights to new areas such as the economic arena, to include ideas like the right to a job, to membership in an organization that represents the interests of workers, and to a standard of living that meets basic needs. These struggles are also addressed to expanding social citizenship, which includes the right to community and to the existence and use of public space. Finally, trade unions and social movements have connected citizenship to a set of communication rights, including the right to communicate (not just to receive information) and to those services that make the right to communicate possible, such as the right to education and to the means of communication, including universal public postal service and universal access to the telephone and now to the Internet. It is worth noting that citizenship may also be used as an exclusive concept, not just an inclusive one: in other words, as a way of dividing "us" from "them." This can be done positively, in order to build internal solidarity, or negatively as a means to exclude people from the benefits of membership in a collective. However, on balance, the struggle over citizenship in the political, economic, and social realm more often represents a struggle for inclusion, focused on the effort to extend the right to fully participate in all areas of society.

One way to think about the importance of these ideas is to consider the critical edge they provide for understanding organized labor in the academic world. Faculty unions have been growing at colleges and universities in both Canada and the United States, particularly in public universities. The fact this growth has occurred during a time of overall decline for organized labor is one of the underreported stories to be told about the contemporary labor movement. However, professors' unions tend to be independent associations, which are only loosely coordinated at a national level by groups like the American Association of University Professors (AAUP) or the Canadian Association of University Teachers (CAUT). They have practically no connections to unions outside the university and college system. In many cases, they have little or no connections to other unions on campus, including unions that represent part-time faculty or graduate teaching assistants who aspire to join the professoriate. This is becoming particularly significant, given the erosion of the tenure system that had been dominant in universities and colleges. According to the AAUP, 48 percent of all faculty members in the United States serve in part-time appointments, and non-tenure-track positions of all types account for 68 percent of all faculty appointments (AAUP 2007). In Canada, the proportion of part-time or contingent faculty members doubled between 1999 and 2005 (Lin 2006). In many respects, professors' unions are craft unions, organized around the mastery of sets of skills and technologies, and around control over information. As such, they are somewhat reminiscent of the older communication-sector craft unions like the International Typographical Union (ITU). As chapter 4 describes, the ITU exhibited qualities that mark many of these professorial unions, including a commitment to democracy and to the view that workers are citizens with public responsibilities. At the same time, the ITU's power base was built on control—over admission to the craft, over training and certification of skill, over allocation of work, and, ultimately, over the operation of the newspaper production process. Today's academic unions lack some of these control mechanisms, but they are characterized by internal democracy, exclusivity, and a limited conception of citizen engagement.

A more complete conception of democracy and citizenship would have us examine whether and how these unions are working with other unions, especially with other knowledge-worker organizations both inside and outside the university, to build a more democratic labor movement, to support democracy in society, and to expand their external activities to make them better citizens of an information society. Despite its power, the ITU was ultimately unable to resist determined campaigns by publishers to use new technologies to erode the union's power base. It was also unable to rally other newspaper unions to its side, in part because of its history of exclusion, of defending its own craft to the detriment of other workers. The remnants of the ITU have been absorbed by the new, converged communication unions. Will today's labor aristocracy wait until its rump is similarly absorbed? Or will it use the privileged position it enjoys to the advantage of the labor movement and of workers worldwide?

Taking this broader view risks departing from the comfort of a simple description of existing data. But it provides an opportunity to take on central issues of our day. Consider a concrete example. In June 2000, AAUP and the Newspaper Guild sector of the CWA held a joint conference titled "Intellectual Workers and Essential Freedoms: Journalists and Academia in the Twenty-first Century." Two years later, in November 2002, the CAUT and the CEP union held a similar joint conference, titled "Disciplining Dissent: The Curbing of Free Expression in Academia and the Media." Both meetings were based on the recognition that professors and media workers are facing the same kind of challenges from employers and the state. By bringing together representatives from both sides, the sessions sought to promote an awareness that academics and media workers need to work together to defend the right to communicate in the face of increasingly hostile governments, corporate concentration, and growing efforts to commodify information and education. Perhaps because of novelty, not much was done once these unprecedented meetings ended. A critical understanding of knowledge labor helps us to think about what else needs to be done, including working together on political and policy issues that range widely to include media concentration; the commercialization of the university; justice for freelance writers, sessional instructors, and other part-time faculty; and the right to control the uses of our work, whether it appears on the printed page, the broadcast airwaves, or the Internet, or in a set of lectures.

This conception of labor citizenship, particularly among workers intensively involved in the use of technology and in communicating messages to students or to broader audiences, is beginning to attract attention among labor scholars. For example, in response to the central question "Why organize?" Paul Johnston has called for reframing the labor movement as a citizenship movement. As business unionism crumbles under the weight of neoliberalism, one positive alternative is to build a movement that will "seek to defend, exercise, and extend the boundaries of citizenship" (Johnston 2001, 35). Concretely, for Johnston, this means envisioning claims and orienting strategies to the status and fate of communities, not just to the benefit of a specific bargaining unit. It also means that labor power should be based on heterogeneous coalitions growing out of a common vision of the future, or a common potential political agenda, and that these visions should not be dependent on market forces. We will take up these issues again, especially in chapter 6, which addresses social movement unionism. For now, it is sufficient to acknowledge that they open a path built on specific theoretical formulations that help to address the question of what happens when knowledge workers succeed or fail to unite.

Chapter Three

Women and Work: Feminism and Political Economy

In *Cat's Cradle*, his dystopian 1963 novel on the dangers of knowledge work, Kurt Vonnegut Jr. takes his readers on a late-December visit to the research laboratories of the General Forge and Foundry Company in the mythical town of Ilium, New York. There we encounter Dr. Asa Breed, the head of the laboratory, who proudly reports that seven hundred people work in the lab. Less than one hundred of them actually do research. The others, Dr. Breed explains, are "housekeepers of some sort" whose labor frees the scientists to play with ideas, like how to restack the atoms in a water molecule to change its freezing point. Dr. Breed calls himself "the chiefest housekeeper of them all," but that title really belongs to Miss Naomi Faust, his longtime secretary. Miss Faust, who is busy hanging Christmas decorations around the office, tells Dr. Breed she has the chocolate bars for the "Girl Pool," the typing bureau in the laboratory's basement. Dr. Breed is relieved; the chore of buying chocolate had slipped his mind. Vonnegut writes:

> All year long . . . the girls in the Girl Pool listened to the faceless voices of scientists on dictaphone records—records brought by the mail girls. Once a year the girls left their cloister of cement block to go a-caroling—to get their chocolate bars from Dr. Asa Breed. "They serve science, too," Dr. Breed testified, "even though they may not understand a word of it. God bless them, every one!" (1963, 34)

Cat's Cradle is a fantastical, darkly comic novel about human arrogance. But Vonnegut's description of the gender relations at the General Forge and Foundry Company lab, where the men work in airy laboratories on big ideas that could destroy the world and the women work in a hierarchy that ranges from mail girl to "Girl Pool" to private secretary, displays a remarkable realism. And it says a lot about the gendering of knowledge work.

The previous chapter addressed a range of ways to define and theorize knowledge labor. Although it gave some attention to the gendering of knowledge work, gender was not a particular focus. This chapter seeks to redress that imbalance. Given the importance of gender in the area of knowledge labor and the relative lack of attention to it in accounts of the development of knowledge-worker theory, we feel it is both useful and necessary to devote a separate chapter to theorizing the gender issue.

The postindustrial society thesis had a significant influence on scholars, policymakers, and the wider society, and it provoked widespread debate. The thesis guided research on the relative decline of the manufacturing sector, the growth of knowledge and service labor, and the varying and changing skills this required. But it did not dwell, for example, on the new range of health hazards—eye strain, repetitive stress injuries, back pain, and other physical and mental health disorders—that cropped up in knowledge-worker jobs that were and still are overwhelmingly staffed by women, especially clerical and service jobs. Furthermore, while there is agreement that an ever-shifting process of deskilling, upskilling, and reskilling is taking place in the occupational hierarchy of knowledge workers, the different ways this affects women and men are often neglected. Women workers, in fact, are likely to suffer the most. They experience deskilling and labor intensification as they are pushed into work that constitutes the peripheral labor market, into jobs that are increasingly found in temporary help industries and in other precarious sectors (Vosko 2000, 2006; Wigfield 2001). Vonnegut's "Girl Pool," so central to knowledge work in the 1950s and 1960s, has disappeared from the landscape. The work once done by the "Girls" has been deskilled through the application of digital technology. The jobs have disappeared, and the remaining "Girls" are most likely working for a temporary help agency or a call center, doing work Buchanan and Koch-Schulte (2004) characterize as a form of precarious, gendered labor that makes extensive use of communication and information technology. Indeed, the call center, which has grown to occupy a central place in the North American economy, is in many ways the contemporary analog to the "Girl Pool" of an earlier generation. Both are places where women work side by side but in isolation from one another, electronically serving the needs of people they never meet in person. But while members of the "Girl Pool" were likely to be permanent employees of the same corporation as the men whose work they typed, call-center jobs, especially in centers that work from contract to contract, may be far less secure.

With the rapid development and expansion of communication technologies and infrastructure around the globe, corporations are empowered to offshore and outsource jobs when deskilling or job elimination is not possible. In today's workplace, where traditional geographical boundaries count for less, workers are promised employability rather than job security, differential salaries pegged to market rates rather than wages tied to length of service, and networking opportunities rather than promotion opportunities (Stone 2004). Furthermore, business policies are creating and destroying jobs at an accelerated pace, resulting in the disruption of individual careers and communities (Fraser 2001; Menzies 1996; Ross 2003). Of course, these trends have different impacts on different workers—hence the significance of considering the gender dimension when analyzing the consequences of globalization and offshoring.

Some research on knowledge workers has included a focus on women, documenting patterns of employment, equity, and wages. Across the knowledge sector, the employment of women continues to grow, but their pay is low outside of a few specific niches. Pay equity continues to be a serious issue, and unions make a big difference. Consider the increasingly integrated communication sector. Women now occupy just under half the jobs in the U.S. communication and media sector, which is similar to the pattern in the U.S. occupational structure overall, and 54 percent of jobs in the fast-growing wireless sector (Lovell, Hartmann, and Koski 2006, 2). They also hold the majority of jobs in the wired side of telecommunications—for the most part the traditional telephone industry. Wired telephony has the highest unionization rate in the communication and media sector and the highest pay rates. As a result, the traditional telephone business has been a source of well-paid, secure jobs for women for many years. Lovell, Hartmann, and Koski report that women earned $623 a week in 2005 in traditional telephony, compared with $573 in wireless telephony (2006, 3). However, relative employment in this area is declining with automation and the shift to wireless. This is especially significant for women. As traditional telephone company jobs disappear, women telephone workers are forced to move into positions with poorer wages and little or no union protection in the wireless telephone or call-center industries. Meanwhile, pay gaps persist throughout the knowledge sector. They are most pronounced in industries with the highest overall pay rates for both men and women. For example, women earn 71 and 72 cents respectively for every dollar men earn in both wired and wireless telecommunications, where wages for both groups are high. By contrast women earn 82 cents for every dollar earned by men in the newspaper business, 84 cents in motion pictures, and 90 cents in broadcasting and cable television. Unions make a big difference. According to Lovell, Hartmann, and Koski, "high rates of unionization are associated with much higher median earnings, both within industries and in specific occupations" (2006, 4). For example, unions represent two out of five workers in wired telephony and union membership adds a full 50 percent to earnings. This is the case in both high-paid and low-paid jobs, including jobs dominated by women such as the growing occupation of customer service representative.

Much of the research on women and work focuses on wage labor outside the home. This chapter, by contrast, takes up a set of conceptual issues across the home-work divide. Specifically, and in the spirit of convergence, we examine and theorize two relationships. The first is the association between feminist theory and political economy; the second between work and home. Meehan and Riordan (2002) have provided a model for how to think about feminism and political economy in the field of information and communication studies. Balka and Smith (2000) make labor a more specific focus in their book on the gendered nature of computerization in the workplace. This chapter builds on this research by demonstrating the value of bringing together feminist and political economic ways of thinking about labor in the broadly defined information, communication, and knowledge arena. Feminist research makes gender, including the gendered division of labor, its starting point, while political economy concentrates on the power relations that mutually constitute the production, distribution, and exchange of resources, including information resources (Mosco 1996). These perspectives are not mutually exclusive; indeed, a primary purpose of this chapter is to demonstrate just how each can enrich and inform the other.

Specifically, this chapter provides four perspectives that are organized around the need to build connections between political economic and feminist research. All four recognize the importance of the communication, information, and knowledge arena as a substantive base for exploring that relationship. Furthermore, the chapter maintains that it is not possible to fully comprehend the relationship of gender to the workplace without understanding the connection between the workplace and the home. That is because for most women over most of recorded history, home and workplace were one and the same. Even when women's work moves out of the home with the development of wage labor, it still bears a distinct relationship to the household. Finally, this chapter suggests that a feminist analysis of labor is a crucial component of research in the field, placing in the forefront Zillah Eisenstein's (1998) argument that global capitalism thrives not only on a transnational division of labor but on a sexual division of labor as well.

Homework as Unpaid Domestic Labor

The first perspective identifies home and work as separate categories of analysis but draws on political economic and labor theory to understand domestic labor. It is important at the outset to identify the home as a nexus for the reproduction and maintenance of labor power. Gender is an inseparable element of this reality because traditionally, women do the housework. A feminist analysis of domestic work and capitalism offers a crucial site of departure for this perspective, one that recognizes the complex structures of work, home, and gender relations (see Gerstein 1979; Rubin 1975). Furthermore, this approach uses the language of wage labor and the market, particularly the processes of commodification and decommodification, to make sense of what has historically been women's unpaid work. Such a perspective adds fresh dimensions to understanding the function of household labor in capitalism, particularly the forms of women's exploitation, and uncovers grounds for feminist resistance.

One impetus for this work is to literally account for women's labor in the home by constructing a feminist economics that starts from the need to calculate the value of women's unpaid labor, enabling us to understand how it contributes to the overall economy. A pioneer in this area is Marilyn Waring, whose 1988 book, *If Women Counted*, produced a detailed critique of the United Nations System of National Accounts (UNSNA). The UNSNA was drawn up with the intention of providing a global snapshot of national economies that could be used by policymakers to make decisions on the exploitation and allocation of resources. But because it uses the tools of neoclassical economics, the determination of income and wealth is based only on wage labor in the marketplace and on the market value of property. Domestic labor does not make the count. This means that homework usually performed by women—such as child care, cooking, cleaning, home maintenance, servicing and repair of household goods, care of the sick or elderly, transportation of household members or their goods, and so on—does not show up in the UNSNA tally. Ironically, these same services are included in the accounting system when paid for privately or provided by government.

Waring's analysis contributed to reforms in some accounting systems, such as the development of so-called satellite accounts that include unpaid work. In addition, her critique helped inspire a range of alternative accounting systems that analyze other forms of value and are not based on the notion that economic growth is "the thing which matters most" (Waring 2003, 37). This research has proven to be especially important, not just in feminist studies or labor studies, but also in fields like environmental studies, which takes into account the value of the natural environment, and in areas of sociology that are concerned with the informal economy that arises out of social networks in physical and virtual communities. But much more needs to be done. According to Waring, "The new feminist challenge is to identify and use these models in public policy making and in advocacy for change" (2003, 35).

Another focus of scholarship on home labor originated in feminist technology studies. Some of the most important work examines the history of household technology. Kleinegger (1987) addressed the transformation of women farmers' work in the first half of the twentieth century, providing an important example of how to integrate a feminist and political economic analysis of domestic labor. She starts by documenting the traditional integration of women's household labor into the farm economy of the nineteenth century. By the early twentieth century, Kleinegger shows, in region after region, much of this production was removed from the home and from the supervisory control of women. As dairy, truck, and poultry farming became organized in large firms engaged in what would later come to be known as agribusiness, another step was taken in the masculinization of agriculture. By the middle of the twentieth century, this regendering of farmwork was embodied in increased specialization, mechanization, scientific expertise, and capital investment. The typical farm woman had little or no access to these activities. Instead, women were returned to the home where they managed the mechanization of the household. This activity did little to liberate women because they "spent at least as much time as before meeting higher standards, and the sexual division of labor was reinforced" (Kleinegger 1987, 178).

Cowan (1979, 1983), studying a series of innovations that amounted to an industrial revolution in the home in the early twentieth century, found that new technologies like electricity and indoor plumbing brought about significant changes in the labor process of the household. Servants—mainly women who were paid, however poorly, for their labor—disappeared from middle-class homes. The number of tasks for which the unpaid housewife was responsible increased as the work became more generalized. This was the opposite of what was happening to men's work in the factory, where industrialization and scientific management led to specialization that divided work into discrete tasks. In addition, while the labor process in the factory tended to reduce the emotional context of men's work, the industrial revolution in the home had the opposite effect on women. Advertisers in popular women's magazines were able to use guilt and embarrassment to prod women into adopting higher standards of cleanliness and household organization, thus creating markets for new cleaning products and household goods. Some contemporary observers were aware of the contradictory nature of the industrial revolution in the home. For example, an article in a farming magazine in 1900 took a critical—and, ultimately, more accurate—view on the impact of so-called laborsaving devices in the home:

> So many labor-saving devices have become common that woman's work should now be simpler, but with these improvements our standard of comfort has been so greatly raised that the present-day housewife seems more overworked. It is indeed unfortunate if we permit labor-saving devices to increase our work, rather than to lessen it. (*Rural New Yorker*, January 20, 1900)

As documented by research in feminist technology studies, the commodification of the home was grounded in material practice. But it was also supported by a culture that had welcomed scientific management into the workplace and proved open to using it to transform the household as well. F. W. Taylor is most closely associated with the approach that called on managers to systematically measure and monitor every aspect of worker activity in order to maximize the efficiency and effectiveness of the workplace. Taylor's approach provided scientific grounding for giving management complete control over the workplace. Specifically, this meant concentrating mental production in those who ran the workplace and leaving the brute physical labor of material production to the workers. For obvious reasons, workers and their organizations opposed what came to be called Taylorism, and the ensuing battles over its implementation are responsible for its decidedly checkered history.

A significant, though less well-known, part of scientific management's history is the attempt to apply its principles in the home and create the most economically efficient household. Christine Frederick, consulting household editor for *Ladies Home Journal* from 1912 to 1919 and the author of numerous books and pamphlets on scientific management in the home, was a leading proponent of this effort. Her 1913 book, *The New Housekeeping*, began by describing her conversion to the "efficiency gospel" under the influence of her husband and a male efficiency expert. Her 1919 book, *Household Engineering*, described using time and motion studies throughout the house to achieve the same efficiencies that the stopwatch was aiming to bring about in the factory. (For example, forty-seven minutes was the ideal time to clean three rooms). Lillian Gilbreth explored similar themes in the 1920s, applying to housework some of the techniques she and her husband Frank developed as efficiency experts studying industry. Her 1927 book, *The Home-Maker and Her Job*, promised women a philosophy to make housework more satisfying, a set of techniques to make it easier, and a method to make it more interesting. As Berch (1980) describes, the movement had a powerful influence on women's identity by associating household management with progressive values just at a time when women were struggling for the opportunity to leave the home.

In *The Making of a Cybertariat* (2003), Ursula Huws builds on the growing body of literature on women, work, and the home with more than two decades of her own research to challenge some of the truisms that limit understanding in the field (see Wright 1987). She observes the persistence of the view that technology saves work time in the home, though instead of promising more leisure it now offers to free women to enter the world of wage labor. There is no doubt that electrification and gas power distribution, which made possible the modern refrigerator, stove, washing machine, and the numerous other appliances that fill most homes in the developed world, cut substantially the amount of time it takes to complete certain tasks. But as

Huws documents, these household technologies create time commitments of their own and contribute to the expansion of needs and desires that take up time in new ways. A considerable amount of time must be spent on the purchase and upkeep of appliances, with further energy invested in making the best use of these items. Purchasing the right stove and fridge with just the right customized features, including the new "smart" panels that suggest recipes based on what's in the fridge or let you know when the milk is about to spoil, is a significant component of creating and running a modern household. Even more significant in eroding the "laborsaving device" thesis is the proliferation of new and time-consuming ways to use home technologies. With the right appliances, everyone can—and, more importantly, everyone should—become a master chef, as well an expert in home decorating, design, and maintenance. These jobs take time, usually the time considered saved by the devices themselves.

In essence, this perspective suggests that capitalism commodifies the domestic environment by promoting the view that the value of what takes place within the home depends on the value of the tools, toys, and products that fill it, and the services its occupants can purchase to maintain or expand it. The list of appliances considered indispensable for a contemporary home in the developed world—and increasingly being marketed in the developing world as well—includes the full range of communication and information technologies: cable and satellite-delivered television, high-speed Internet services, multiple wired and wireless telephones, high-end video and sound systems, digital cameras and video recorders, and all of the customized add-ons that are considered essential to realize the full potential of these devices and services. The all-electric home promoted by the General Electric Company in the 1950s—including the one it built for Ronald and Nancy Reagan for use in GE advertisements—appeared futuristic, but it barely hinted at the extent of the communication, information, and entertainment systems that would fill households fifty years later. It also neglected to note that operating and managing all of these new systems would require substantial labor commitments in terms of both time and skill.

Scholars who study the economic history of the household have found that the growth of women's wages in the formal economy increased pressure on labor productivity in the home. Why should a woman stay in an unproductive home when she could earn a living, however modest, in the marketplace? Concurrently, rising men's wages increased the demand for goods and services in the home. A decline in household servants put pressure on higher-income households to incorporate replacement technologies like dishwashers or microwave ovens. Finally, the expansion of media, especially advertising, helped to create a consumer culture containing both information about and pressure to adopt new technologies (Day 1992). There has always been resistance to the growth of technology in the home. Ours is not the only time when people experience the stress of creating and maintaining the perfect household, or of keeping up with new developments that may or may not satisfy needs. But the prospect of achieving some version of the perfect household has always nudged people, sometimes reluctantly, sometimes enthusiastically, to add new technologies to their homes.

The expansion of capitalism into the home has not, however, resulted in any substantial advance in bringing household work into the orbit of wage labor. Capitalism expands its presence in the household by generating new technologies and the new needs they help to create. But it stops short of paying those who work on them for their labor. And although men have become increasingly involved in areas like communication technology management, women are still largely responsible for much of the household management process. It is women, therefore, whose exploitation in the home is deepened. Research on time spent on household management indicates that men enjoy at least thirty minutes more of free time per day than women (Mattingly and Bianchi 2003). Furthermore, since the modern household makes enormous financial demands for capital investment in technology and in financing the operation of household machinery and the services it provides, there are strong pressures to turn unpaid household labor into paid nonhousehold labor, either in the home or outside it. But that does not necessarily make women's lives easier, or happier. A recent study shows that since the 1960s, men have gradually cut back on activities they find unpleasant. As a result, they work less and relax more. Women, by contrast, have replaced housework with paid work and have not been able to cut back on activities they find unpleasant. A typical woman in the 1960s spent twenty-three hours a week in activities she considered unpleasant, or forty more minutes than a typical man. Today the gap is ninety minutes (Leonhardt 2007b).

What start out as conceptually separate arenas—home and workplace—are therefore more properly seen as zones of mutual constitution. As the need grows for wage labor to maintain the home, women (and men) whose household labor goes unrewarded in the wage system are either bringing wage labor into the home, or going out into the traditional workplace. A third possibility would be to mobilize the labor power of unwaged household laborers. Efforts in this area have not been particularly successful, however. Nevertheless, Huws maintains that labor organizations and feminist social movements need to make this a priority because the growth of what amounts to labor-creating devices in the home will only deepen the problem of unwaged household labor in the future. Moreover, some analysts argue that the expansion of women's unpaid labor is symptomatic of a deeper extension of labor into the consumer realm, as those who purchase goods and services are increasingly forced to contribute to the labor required to supply them. This so-called voluntary labor means that, among many other things, we now pump our own gas, clear our own tables in fast food restaurants, dial our own telephone calls, assemble our own furniture, or find and bag our own groceries in supermarkets (Fox 2002).

Scholars studying communication and information technology have made important contributions to documenting the nature of unpaid work in the home. Years ago, the political economist of communication Dallas Smythe established a groundwork for this kind of research by describing the extent of audience labor in the home through the sale of people's attention to advertisers. Smythe (1977) calculated the capital that families invested in buying and operating the equipment—at the time it was mainly broadcast radio and television receivers—and the value of the time audiences spent delivering their attention to the advertisers who sponsored the programs the families listened to and watched. Smythe concluded that unwaged audience labor propelled the commercial broadcasting system. However challenging the task, Smythe argued that it would be as essential to mobilize audience labor, including that of unwaged women in the home, as it had been to bring together those working in offices and factories. Meehan (2002) built on this work to examine the commodification of audiences through the broadcast ratings services. She has also extended it by constructing an important bridge between Smythe's political economic analysis and feminist research.

Drawing on this work and expanding its conceptual grounding with the research conducted by Huws and others could allow us to calculate the total wage bill for uncompensated household labor. Combine this with the absence of many of the benefits that typically accompany waged work, including health insurance, disability insurance, pension plans, and so on, and we would likely come up with a staggering sum that would serve as one index of the enormous gap between what the labor, feminist, and other social movements have achieved and what is left to be done.

Homework as Wage Labor

The previous section focused on the penetration of capitalism into the home and how it profits from unwaged or free labor. We now turn our attention to the expansion of the workplace into the home, turning it into a site for traditional wage labor. The contemporary incarnation of global capitalism creates a multiplicity of roles for women, as wives, housekeepers, child rearers, *and* low-paid workers (Eisenstein 1998). The latter role has been documented at the base of the Silicon Valley food chain since the 1980s, when immigrant women began to be employed in hardware production and assembly in factories and in their homes. As Hossfeld (1990, 155) reports, there were significant tensions around gender, race, and class right from the beginning, as mainly white, male, middle-class managers confronted workers who were almost universally poor women from Hispanic, African American, and Asian backgrounds. It is worth noting that although exploitation, including long hours and dangerous conditions, was an enduring force, so too was resistance, as women used numerous techniques to fight back.

Examining high-tech work at the turn of this century, one especially valuable exemplar is Pellow and Park's 2002 book, *The Silicon Valley of Dreams*. It documents the expansion into the home of an industry that relies on low-wage female labor, which is often immigrant labor as well, to carry out basic electronics assembly work. Here, homework is not unpaid domestic labor; rather, it is the wage labor of women with few alternatives than to turn their homes into extensions of factories or call centers (Head 2003).

Pellow and Park document the dark side of Silicon Valley, the side that has received little attention in the hype about the region as a magical realm of clean work, the capital of what Bill Gates (1996) has called "friction-free capitalism." Certainly, Silicon Valley, the nickname for the concentration of high-tech businesses that occupy California's Santa Clara County, contains neighborhoods that are models for recycling, cleanliness, and environmentally sensitive urban design. Meanwhile, its many computer companies operate clean rooms to keep out the impurities that can destroy an expensive fabrication process. But Pellow and Park show that Silicon Valley is also an environmental disaster area, with more of the most toxic or "superfund" sites in the United States than in any other location (2002, 76). Santa Clara County is littered with chemical dumps for the highly toxic materials that are essential in the manufacture of computers and peripheral devices. The dirty and dangerous neighborhoods near the dumps, largely occupied by poor people and immigrants, are the ones Pellow and Park single out for attention.

Environmental protection is very much a gender issue in the valley. It is common practice for women to work at home as "independent contractors" producing computer components. This means they bring into their homes the same toxic chemicals that fill the dumps that require special government appropriations for a cleanup. But none of this money goes to clean up the kitchens of the women who brew up a dangerous chemical soup. Why do women do this kind of work at home? For some, it is the only way to bring in money and take care of children at the same time. For others, it means the ability to earn money outside the watchful eyes of the U.S. Citizenship and Immigration Service. One of the valley's dirty little secrets is the use of low-wage immigrant labor, including the labor of illegal immigrants, to carry out some of its essential work (Pellowand Park 2002, 158–63).

There is nothing new about homework, including industry reliance on newly arrived legal or illegal immigrants. Decades ago, the U.S. textile industry was heavily dependent on homework. After the Second World War, the U.S. government banned industrial homework in seven industries—including women's apparel, embroideries, handkerchief manufacturing, jewelry manufacturing, and knitted outerwear—in response to evidence that minimum-wage and child-labor laws were routinely being violated. Most of the bans were lifted in 1989, with the exception of women's apparel and "unsafe" jewelry production (Edwards and Field-Hendrey 2002, 172). Since then, resistance to exploitation in waged household labor has been stepped up. Research has documented organizing activities by traditional trade unions and the creation of nonunion worker associations, both of which push for better pay and working conditions for the people Louie (2001) has called today's "sweatshop warriors."

What makes the work that Pellow and Park describe particularly significant is that it is an important dimension of what is arguably the most strategically significant industry in the world, one that promises a new grounding for work: information rich, environmentally sound, and likely to lead from traditional capitalism to a postindustrial (Bell 1973) or networked (Castells 2000) capitalist society. But for those working at the bottom of the knowledge society food chain, forms of exploitation continue to turn the home into a site of production, and sometimes toxicity.

According to Pellow and Park, thousands of immigrants, including Cambodians, Koreans, Vietnamese, Latinos, and Filipinos, work on production jobs out of their homes in Silicon Valley. While much of the attention in the homework field has gone to white-collar workers who do contract jobs writing software and managing websites (Barley and Kunda 2004), the women homeworkers who make cables, circuit boards, and other computer components do so in their kitchens, bedrooms, bathrooms, living rooms, basements, or garages. They routinely work with acids, lead, and solder, a chemical mix essential to construct modern communication infrastructures. Because they are paid at piece rates as low as a penny per component, it is not unusual for them to get their children, parents, or other extended family members to help out. These women work long hours under the constant threat of dismissal (Pellow and Park 2002, 158–60). Health problems are common, ranging from neck and back pain through eye strain and sleep deprivation to respiratory disorders, problem pregnancies, and the inevitable serious health disorders like cancer brought about by long-term exposure to toxic substances. As research on the consequences of the high-tech industry's "dirty little secret" circulate in various media accounts, especially in the independent media, efforts to organize paid home workers have grown, particularly through social movement labor organizations. These are examined more thoroughly in chapter 6, but it is important to observe here that this has led to some rethinking of labor organizing strategies in the electronics industry. This includes a recognition of the need to pursue a global strategy that takes into account the expanding international division of labor in the industry, as well as the need to build ties to consumer and other social movement organizations that operate in the household and industrial electronics sectors (Smith, Sonnenfeld, and Pellow 2006, esp. 191–99).

In recent years, communication scholars have focused their attention on new media, including the worldwide spread of the Internet, cellular phones, and other computer communication systems. A number of scholars have looked at how these technologies enable women to do paid work in—or from—their homes, through so-called telework arrangements that use telecommunications to replace the daily commute to a central office. For example, telework has been presented as an option that allows workers, especially women, to achieve a better balance between their work and home lives. Surveys on why women pursue telework in the first place indicate that the ability to spend more time with children by combining work with child rearing is at the top of the list (Beasley, Lomo-David, and Seubert 2001, 479). However, some studies have suggested that rather than allowing women to improve the balance between the demands of work and family, telework increases the strains (Duxbury and Higgins 2001; Menzies 1996; Tremblay, Paquet, and Najem 2006). In addition, Tremblay, Paquet, and Najem have raised serious questions about who benefits from telework, and why such arrangements are put in place. They examined data from Statistics Canada's Workplace and Employee Survey—covering six thousand firms and twenty-three thousand workers between 1999 and 2002—to assess how workers feel about telework. They found that two-thirds of teleworkers take on the job of working at home for all or part of the week, not because they want to, but because their employer demands it. These studies suggest that there is much less actual benefit in using telework to overcome the imbalance in work and family life than earlier commentators predicted.

The use of new media in homework is an important development and one that is well worth further study. At the same time, however, it is clear that scholars have not probed deeply enough into the core of the production process that makes these technologies possible. As a result, very little work addresses the role of homework in building the information infrastructure. Given that the vast majority of homeworkers are women, this area should be of particular concern to those interested in a feminist analysis of communication and information technologies. Moreover, this research need not only focus on the admittedly significant problem of exploitation. One of the more remarkable stories to come out of Silicon Valley and other centers of information-technology production in the developed and developing world is the growth of resistance to exploitation, particularly through the formation of unions, labor associations, and community-based social movements to bring change to the industry (Louie 2001; McLaughlin and Johnson 2007; Smith, Sonnenfeld, and Pellow 2006).

Dissolving the Divide between Work and Home

Case studies documenting the history of the relationship between women's work and technological change maintain that it is very difficult to generalize on the microlevel, or at the point of production. Some go so far as to suggest that general conclusions can only be drawn within specific industries and technologies (de Groot and Schrover 1995, 1). Others argue that studies documenting deskilling and worker alienation in women's work neglect the general pleasure that is associated with working with new technology, such as the speed or flexibility the devices allow. In effect, the technological sublime described by historians Leo Marx (1964) and David Nye (1994) provides a cultural orientation that often enables a more pleasurable response than we might expect by looking only at the material labor process (Korvajarvi 2003).

It is at this point that a third perspective on work and home intervenes in the debate. This one sees the erosion of both categories, as home and work become so interpenetrated that they lose what independent value remains. Best exemplified in the work of the autonomous school of scholars, including Antonio Negri and Tiziana Terranova, this perspective concentrates on the emancipatory potential of new technologies and the accompanying rise of a knowledge-worker class. In this point of view, the home and the traditional workplace both become places of work and play, mutually constituted and a fundamental challenge to capitalist efforts to reduce all activity to wage labor. Building on a long tradition of Italian social theory, Hardt and Negri gave this autonomist perspective wide circulation in their 2000 book *Empire*. A central pillar of the work is a belief in the rise of immaterial labor, another term for what scholars, primarily in North America, would call knowledge work (Huws 2006).

Although the center of autonomist theory is Italy, it has enjoyed a base in the United States for quite some time. One of the most important North American autonomists, Harry Cleaver, gained prominence by addressing its relationship to Marxist theory, and to labor and gender as well. In a 1993 interview with the Italian journal *vis-a-vis* he took up these themes. For Cleaver (1993), "autonomist" refers to the power of workers to define themselves autonomously "from capital, from their official organizations (e.g., the trade unions, the political parties) and, indeed, the power of particular groups of workers to act autonomously from other groups (e.g., women from men)." The term also represents a recognition of people's right to struggle against their reduction to "mere worker." In the interview, Cleaver said: "Precisely because capital seeks to intervene and shape all of life, all of life rebels, each nook and cranny of it becomes a site of insurgency against this subordination. Housewives go on strike in the home or march out of it collectively into the streets." Gender takes a central place in this vision, not only because the home becomes a site of resistance. Autonomy also means a refusal and a reconstitution, for "much of the revolt of women can be seen as a refusal of their traditional roles in the social factory: as procreators and recreators of labor power accompanied by demands for new kinds of gender and other social relations."

If autonomist theory was not born with the knowledge economy, its development is certainly closely associated with the rise of digital technology. Autonomist scholars have argued that the digital economy is qualitatively different from one based on manufacturing because it requires a qualitatively different kind of work. Specifically, because it demands a great deal of skill and creativity, it is much more difficult to manage and control (Dowling, Nunes, and Trott 2007; Terranova 2000, 2004; see *Ephemera*, February 2007). As such, these scholars have sought to uncover the transformational and libratory potential of the new forms of work that are prominent in the knowledge economy. The autonomists have accepted many of the critical positions taken by those who attack all forms of exploitation in the workplace, but they take issue with the view that knowledge work is similar enough to earlier forms of labor (agricultural and industrial) that there is little potential outside of traditional trade unionist forms of resistance to overcome or ameliorate exploitation. For example, Terranova acknowledges the plight of those she calls "NetSlaves," primarily skilled workers in the software industry who work long hours in insecure jobs (2000, 33). But she chooses to focus on what she sees as a qualitatively different kind of work in the digital economy: mainly creative labor that resists attempts to apply the Taylorist principles of control that were effective in the industrial era. From her perspective, and that of other autonomists like Lazzarato (1997), it is very difficult to capture this type of creative labor because it is such a challenge for capital to routinize it. Such work can—and often does—take place virtually anywhere, from the traditional workplace to the home, and even in wireless-enabled coffee shops that meet the needs of an increasingly mobile workforce. But it bears the mark of none of these places and, in the process, challenges traditional notions of all of them. Today's labor is neither workplace work nor homework, but represents, as the autonomists like to put it, the creation of a "social factory," or the transition of work processes from factory and office to all of society. It is no longer the case that labor takes place only at work and that play is limited to the home or to places traditionally associated with entertainment and leisure. Accelerated by a digital economy, work and play have become increasingly placeless. This calls into question the very nature of both work and home.

Because of the approaches taken by autonomist scholars that emphasize the "newness" of both contemporary economic formations and knowledge work itself, critics have argued that autonomists constitute the immaterial laborer as a new vanguard subject with the potential for radical social change (Dowling, Nunes, and Trott 2007; Trott 2007; see Harman 2003). This condition, they add, has yet to be generated in reality. In addition, there is evidence to suggest that knowledge workers who perform their work at home and elsewhere are still subject to the managerial gaze, through sophisticated networks of surveillance and screening techniques (Felstead, Jewson, and Walters 2003; Halford 2005). While there are limitations on how far deskilling and routinization can go, management practice and theory have attempted to bring knowledge work into the sphere of control (Davenport 2005; Liker 2004). The autonomists argue, however, that this is less and less effective, partly because workers use new media to carry out resistance through self-organization and the construction of alternative political visions and practices (Taras, Bennett, and Townsend 2004).

Autonomists also see the rise of "consumer labor," which others would criticize as a massive shift from paid to unpaid labor that exploits both worker and consumer, as evidence that the system of paid or waged labor is breaking down beyond the ability of the traditional market system to repair it (Fox 2002). The social factory undermines both work and home, weakening, if not rendering irrelevant, the categories that governed industrial society. From this vantage point, the exploitation that takes place in the digital economy is a vestige of an older era, one that attracts our attention simply because we know it so well. As a result, Terranova insists (2004), we lose sight of the transformation taking place in labor and, by extension, in the structures of everyday life.

Much of this perspective resembles the extension into the work world of Donna Haraway's now-classic *Simians, Cyborgs and Women* (1991), which called on us to question the attempt to completely capture computer communication, and technology in general, within an industrial vision of capitalism, social class, and exploitation. The cyborg could also help create a path to freedom by exploding traditional, repressive categorizations that separate human from machine and human from animal. Rethinking these relationships has been fundamental to Haraway's work. The autonomists extend this to labor by pointing out how the old definitions and conceptions that separate work and home, waged labor and voluntary labor, enjoyment and exploitation, are increasingly inadequate to understanding the digital world.

The idea that computer communication technology can support liberation movements as well as the march of capital has been introduced to communication studies primarily through the work of Nick Dyer-Witheford, whose *Cyber-Marx* (1999) demonstrates the power of what Marx called the "general intellect" in challenging capitalist hegemony. This perspective has found resonance in social movement and social class research, but less so in the feminist literature. A notable exception is the work of Mariarosa Dalla Costa on women and labor (Dalla Costa and Dalla Costa 1999). In one sense, this is surprising because the message of questioning fundamental, especially industrial, categories is certainly a pillar of feminist perspectives. At the same time, however, it is clear that many of the jobs the autonomists describe are held by young men, whose work is to create video games, for example, or the latest in online advertising. The autonomous perspective is less applicable, at least for now, to women working in the knowledge industries. Women's work tends to cluster at the low end of hardware production, as described by Pellow and Park (2002), or in rapidly growing service fields like the call center. Women held more than 63 percent of call center jobs in Canada in 2004 (Akyeampong 2005) and call-centers now employ an estimated 6 million Americans, most of them women. These jobs are more easily industrialized and subjected to scientific management than the creative jobs in the software industry.

In this respect, a feminist reading of autonomist theory applied to the contemporary workplace would see its value primarily in pointing to a gender divide in the kinds of work performed in the digital economy. The idea of "boys working on the toys" best fits an autonomist vision of immaterial labor smashing traditional categories (Kline, Dyer-Witheford, and de Peuter 2003). Meanwhile, the girls are still working in a painfully material world, a world that promises flexibility but delivers precariousness, and that, for many, offers little more than what Leah Vosko (2000) calls "halfway houses for housewives." The autonomist vision holds out the potential for another world. But when the current world of women's work is situated within that alternative, it is clear that bringing the two worlds together is a prodigious, if not impossible, task.

At Home in the House of Labor

The final perspective takes up responses from workers' organizations to the changing configuration of home and work. In essence, it focuses on the workplace, whether in the traditional factory, the office, or the home. But it does so from the perspective of trade unions, labor federations, and social movements that are attempting to deal with the crisis in organized labor.

Interest in a feminist trade union agenda grew in the 1970s, 1980s, and 1990s as women entered the workforce in large numbers. In 1995, the action platform of the United Nations Fourth World Conference on Women recognized the important role of trade unions in promoting gender equality and protecting vulnerable women workers. A few years later, the International Labour Organization (ILO) adopted a declaration of rights and principles at work that called for, among other things, the elimination of various forms of discrimination, including on the basis of sex (ILO 2001). At the same time, as large numbers of women were entering the workforce, trade union leaders were beginning to cope with the decline in the power and influence of labor organizations (Briskin and McDermott 1993; Lawrence 1994). This helped contribute to a growing recognition that new times required a new agenda, one that would incorporate the needs and concerns of the growing proportion of workers who were women. There are different views on what constitutes such an agenda, but there is consensus on three areas. One is recruitment, seen in attempts to bring more women into the labor movement. The second moves the idea of promoting gender equality through collective bargaining into the mainstream of organized labor. This is achieved through efforts to redress existing examples of pay inequity, the creation of clauses on equal pay for work of equal value, policies against sexual harassment, or family-friendly policies such as parental leave and child-care benefits. Finally, individual unions and labor federations are attempting to increase female representation in leadership positions, such as on the union executive or on collective bargaining committees. This is seen as an important way to enhance credibility with both current and prospective female members. Some unions have made significant progress in these areas. Munro (2001) describes the efforts of UNISON, Britain's largest trade union, to incorporate elements of a feminist agenda by including proportionality and fair representation in its constitution, as well as by advancing affirmative action programs to improve job opportunities for women and to increase the numbers of women working for unions. She concludes that the union has a long way to go, especially in advancing the interests of minority and working-class women. Nonetheless, the fact that it continues to make a strong effort in this direction is important to its members, and can also serve as a model for other unions.

The growth of convergent trade unions in the communication and information sector in North America has offered some new opportunities for organized labor to contest important issues, including pay equity, with conglomerate corporations. Nevertheless, these unions tend to be dominated by men, not just in terms of hierarchy but in organizational culture as well. At the same time, declines in traditional union membership and the notorious difficulty of organizing workers, especially women in precarious occupations, have prompted the rise of labor associations or social movement organizations formed to defend the rights of workers by drawing into the movement people who cannot (or will not) join a traditional trade union. Associations like WashTech or Working Today serve men and women information workers, including programmers and content producers. Meanwhile, broader associations like Working America tap directly into the home for support and to make room for women. At the international level, the Union Network International (UNI) has identified the importance of paying attention to the gender dimension of the global labor force and the role this plays in a new approach to unionism. Increasing the number of women who hold executive posts within its member unions has been a priority. This section of the chapter concludes by examining how both groups—traditional labor unions and nontraditional workers' organizations—address the needs and desires of their female membership and how they are providing an opportunity for working women to enter the house of labor.

In order to understand the situation in traditional unions, it is useful to review the growth of convergent unionism. The term refers to efforts to bring unions together into new configurations that are designed to increase labor power horizontally, vertically, or by taking advantage of potential synergies among previously independent groups or workers. Britain's UNISON is an example of convergence, particularly of a movement to bring together service workers, mainly public service employees, under one union banner. Founded in 1993, it has 1.3 million members working full- or part-time in local authorities, the National Health Service, the police service, colleges and schools, the electricity, gas, and water industries, transport, and the voluntary sector. Convergence provides the resources to extend the union's agenda beyond its traditional focus. This is significant for women because it opens up opportunities to put a feminist agenda into practice.

In the North American communication and information sector, a range of media unions—the International Typographical Union (ITU), the Newspaper Guild, and the National Association of Broadcast Employees and Technicians (NABET)—have joined the Communications Workers of America (CWA), founded mainly as a telephone workers union. Calling itself "the union for the information age," the CWA is a model of convergence, representing seven hundred thousand workers employed in telecommunications, broadcasting, cable TV, newspaper and wire service journalism, publishing, electronics, and general manufacturing, as well as airline customer service, government service, health care, education, and other fields. Canada's Communications, Energy and Paperworkers Union (CEP) has pursued a similar strategy. It was founded in a three-way merger in the early 1990s. Since then, it has welcomed Canadian NABET and many Canadian locals from the ITU, the Newspaper Guild, and, most recently, the Graphic Communications International Union (GCIU), whose U.S. parent union joined the International Brotherhood of Teamsters in 2005. CEP's 150,000 members work in pulp and paper mills, telephone and telecommunications companies, newspapers, radio, and television. They are also employed as graphic artists, hotel workers, computer programmers, truck drivers, and nurses.

The diversity of the membership in these converged unions in the United States and Canada is also embodied in gender diversity. Predominantly female occupations like telephone operator or newspaper clerical worker are now covered by the same unions that represent computer programmers and pulp and paper workers. This does not mean to suggest, however, that some sort of even gender distribution prevails throughout the union as a whole. Some parts of these converged unions, such as the NABET or ITU sectors of the CWA, are overwhelmingly male. Others, such as the Newspaper Guild sector, are mixed. Still others, such as the growing customer service cluster in the CWA, are overwhelmingly female.

In the communications businesses, the forces of convergence are often counterbalanced by pressures toward deconvergence. The same is true in organized labor. For instance, a group of seven unions led by the Service Employees International Union (SEIU) pulled out of the AFL-CIO in July 2005 following a long-standing dispute over where the labor federation should concentrate it resources. The SEIU wanted the federation to consolidate its member unions into fewer but larger and more efficient units, to shift money away from political activity and research, and toward grassroots organizing, especially in the service occupations where women predominate. With 1.9 million members, the SEIU was the fastest-growing union in the AFL-CIO and the most diverse: 56 percent of its members are women and 40 percent are people of color (SEIU n.d.-a). The breakaway unions formed a new labor federation, Change to Win, which elected Anna Burger as chair, making her the first woman to serve as leader of a major labor federation in the United States. It also pledged the most aggressive organizing campaign in fifty years, concentrating in industries that increasingly employ women and minorities. Whether Change to Win will succeed remains to be seen, but what is clear is that at the highest levels of organized labor in the United States, there is lively debate over how to address the overall crisis in trade unionism and the specific need to overcome the gender divide that has historically marred the trade union movement.

Critics argue that labor union convergence increases the centralization of power and bureaucracy, thereby making it less likely that union leadership can maintain close contact with the rank-and-file membership. Specifically, this means a growing distance between the predominantly male leadership and female rank-and-file members. Women account for about 40 percent of union members in the United States and close to 50 percent in Canada (Akyeampong

2004; Granville 1999). In October 2005, the forty-six-member Executive Council of the AFL-CIO had thirty-seven men. Women fare better in the Canadian Labour Congress, holding about one-third of the positions on the federation's Executive Council. Culture and unclear union constitutions have also been cited as some of the factors preventing women from actively participating in trade unionism. Joyce Nonde, the first female vice president of UNI-Africa, remarked that it is difficult to implement policies specifically oriented toward women if union structures and responsibilities do not make such initiatives a priority (UNI 2004a). Individual unions, including the CEP and the CWA, have sought to increase the diversity of their governing boards by setting aside specific seats to be held by women or members of visible minorities.

Supporters of labor convergence argue that it gives unions greater clout in collective bargaining, thereby diminishing the relative power that has been concentrated in big companies over the past three decades. This holds the potential for preserving and extending to women and minorities the so-called union wage premium—the difference between what a unionized and a nonunionized worker earns when all other factors are constant. That "premium" averages about 15 percent in Canada. But union membership is especially significant for women, since unions play a major role in closing the wage gap between men and women. Nonunion women make about 76.5 percent as much as nonunion men. Union women, by contrast, make 91 percent as much as union men. The wage gap between the earnings of women who belong to a union and those who do not is even more shocking: a whopping 39 percent (Jackson 2003, 7–9). Therefore, the stronger the union movement, supporters of labor convergence argue, the better it is for women workers. Moreover, they argue that larger unions are able to be more involved in social and political activities, within the workplace and beyond. For example, in 2006 the CEP finally won its long and costly battle against Bell Canada for pay equity for women workers (Freeze 2006). The case began when the union filed a pay equity claim with the Canadian Human Rights Commission in 1992 on behalf of telephone operators and dining service and house service workers, and it went all the way to the Supreme Court of Canada. Fourteen years after it began, the case was settled by mediation. Bell Canada Enterprises agreed to pay up to $100 million to almost five thousand women who worked for the company during the 1990s. Only a strong union could afford to continue such an action against Canada's largest and arguably most powerful conglomerate. Another example, also in Canada, resulted from a 2006 study of freelance writers. It found that overall income was declining, not rising, and that women freelance writers earned $3,000 a year less than men (PWAC 2006, 28). A better deal for freelancers is not just a professional issue, therefore, but a women's issue as well. CEP, working with the Professional Writers Association of Canada

(PWAC), founded the Canadian Freelance Union (CFU), a CEP local and the first of its kind in the country. By 2007, it had attracted about 350 members and was waging a coordinated campaign against a new freelance contract drawn up by the Quebecor media giant. The union urged individual writers not to sign it, and to remind editors that the local was willing to negotiate with Quebecor on a contract they would be willing to sign (Canadian Freelance Union 2007). But collective bargaining is only part of the new local's agenda; it also fights for higher freelance rates, protection of intellectual property, and better benefits for freelancers.

Nevertheless, it is uncertain whether converged unions are genuinely bringing together different kinds of workers in the knowledge, information, and communication sectors, such as print or broadcast reporters and telephone operators, or whether they are merely federations of what are, in effect, dissimilar labor organizations representing dissimilar employees. This is particularly significant for women because the merged unions are typically led not just by men, but by men from merger partners that historically were led by male workers. Neither the CWA nor the CEP, for example, has ever had a woman as president, though their media workers' sections have been led by women: Linda Foley has been president of the CWA Newspaper Guild sector since the mid-1990s and Gail Lem was CEP's national vice president of media from 1996 to 2000. In 2007, two of the top three CWA executives and all three CEP national executive officeholders were men.

For convergence to genuinely benefit women workers, real integration is essential. In the early 1980s Ellen Cassedy and Karen Nussbaum recognized unionization as a means for women to overcome class and gender inequalities in the workplace, notably sexual harassment and discrimination. Their influential 1983 volume *9 to 5: The Working Woman's Guide to Office Survival* charted the multiple challenges facing women in the office, and offered readers a hope that organizing, working collectively, and gaining knowledge about fundamental legal rights in the workplace could provide a sense of empowerment. Today, the National Association of Working Women, which Nussbaum founded, continues to pursue the same goals for women, setting its sights on social justice unionism not simply economic gains.

Social movement unionism has taken a crucial global step within UNI, a Swiss-based organization formed in 2000 from the merger of four union federations spanning commerce, finance, telecommunications, and media. With more than 15 million members in 150 countries, UNI has the potential to foster internationalism like no other labor organization. Part of that role is to tackle dire problems that women face worldwide, both in and out of the workplace: poverty, segregated career paths, discrimination, HIV/AIDS, and the traffic in women for sexual and labor exploitation. Indeed, in all countries, women workers are among those most likely to suffer from the negative impacts of globalization. Occupations dominated by women in the telecommunications, finance, graphical, and other service industries are increasingly suffering from jobs cuts, contracting out, and outsourcing to lower-wage countries or regions (UNI 2005e). In some cases, work that was done mainly by women in the country doing the outsourcing becomes men's work in the country that gains the jobs. For example, women hold between 60 and 70 percent of call-center jobs in North America. In India, where the business is booming, it's the reverse: men hold 60 to 70 percent of the jobs in call centers (Kapur 2007). Whether that pattern will continue in the long run remains to be seen.

UNI World Women's president Barbara Easterling has set a clear trajectory for women in the union movement: "It is women who are the vanguard of a new worldwide labor movement, one that has equality and justice at its core . . . [and] one that is global in its outlook . . . creative in its strategies . . . and tenacious in its tactics" (UNI 2006a). At the 2005 UNI World Women's Conference in Chicago, an action program was forwarded that emphasized organizing women into unions in both the formal and informal economy, in customer services, and in countries receiving outsourced work (UNI 2005e). Some months before, French telecom unions jointly signed an agreement designed to pressure telecommunications companies to reduce the imbalances between men and women in that industry (UNI 2005g). Since its inception, UNI and some affiliate locals have been addressing inequalities between men and women in the workforce, acknowledging what Vosko (2006) defines as the gendered distribution of forms of employment across industries. As a UNI statement entitled *Women and the Information Society* points out, working women are to be found in lower-paid jobs at the bottom of the occupational scale in almost every sector, such as call handling, electronics assembly, teleworking, data entry, and customer service (UNI 2003b). Devoid of job security, income security, and social protections, workers in these occupations are certainly in need of organizing. Importantly, structural changes and union democracy can offer equity-seeking groups and workers in precarious employment greater opportunities to participate fully in the movement (Gupta 2006).

A second response to the crisis in North American labor has been the formation of worker associations or worker movements that provide tangible and intangible benefits without formally negotiating collective agreements. These have been especially prominent in the high-tech sector, where union organizing has been especially difficult. Nevertheless, with increasing numbers of workers, especially women, working full-time but on hourly contracts that contain practically no benefits or overtime pay, developing some form of worker protection in this sector is particularly important. Two types of worker associations feature prominently in the knowledge sector: those that bring together technology-intensive workers, and those that represent workers who primarily produce content. Another type of association—the broader, grassroots organization aimed at bringing workers into coalitions to fight for social change—looks beyond the knowledge sector for members but relies on knowledge workers and communications technology to build membership. The AFL-CIO recently created a community-level organization known as Working America and signed up close to a million members across the United States in less than a year and a half. Working America members agree to pay an annual fee and pledge to cooperate with unions in political and legislative campaigns. Significantly, the organization reaches into the home, the traditional site of women's labor, to target nonunion members and retirees. It does so through door-to-door campaigns and through the Internet. It is especially noteworthy that Working America's founding director was Karen Nussbaum of the original 9 to 5 movement, who has been working on women's and labor issues for more than two decades. According to Nussbaum, "Working America has found a way to harness the power of workers who do not have a union on the job in the fight for better social and economic policies, and corporate and political accountability." The goal is to encourage activism and, eventually, a progressive majority in the United States (Nussbaum 2005).

As with trade union convergence, there is disagreement over the likely success of worker associations in responding to the crisis facing organized labor. Supporters see these groups as a new form of unionism, one that makes use of new technology to reach workers, particularly women, who have little experience with unions (Greene and Kirton 2003). They are also a way of bringing into the labor movement people who, for whatever reason, do not necessarily want to be part of a trade union. Where supporters see worker associations as a start toward rebuilding the labor movement, or perhaps even as a revival of the social movement unionism of earlier years that brought about significant reforms, and a step toward providing genuine representation and leadership roles for women, critics see them as little more than a false hope for women workers. They argue that women need to gain real power, including power to negotiate their wages, and that social movement unions simply cannot deliver it. Even worse, these organizations risk treating women as supporting players for the "real" trade unions.

The Gendering of a Postindustrial Society

With its focus on the shift from manufacturing to informational labor and what this means for the power of knowledge workers, the debate about the shape of postindustrial society has directed remarkably little attention to the gendered nature of informational capitalism or the information society. Nonetheless, empirical studies have demonstrated the specific ways in which knowledge labor is gendered, and have documented its significance. Women are filling jobs in the knowledge sector, but primarily at lower levels and at lower pay than men. Women suffer more workplace injustice—and in more ways—than their male knowledge-worker counterparts. Pay equity is an issue for women at practically all points along the hierarchy of knowledge work. It is an issue for telephone operators as well as university professors, for librarians and for corporate executives as well. Women are more likely to suffer precariousness than men, and their jobs are more likely to be outsourced. According to the ILO's gender promotion program (2001), increasing female labor force participation has not meant improved working conditions or social protection. Misperceptions persist about the role and contributions of women in employment. Women are more likely than men to be victims of labor market failures and distortions. They account for the bulk of atypical workers who are inadequately covered by labor legislation. And they continue to bear the brunt of combining work and family responsibilities.

At the same time, Eisenstein notes (2004), the globalized, digital capitalism that exploits women also provokes and makes visible its activism across the globe. This allows women to build connections to each other that would have been far more difficult in an earlier age:

> As former divisions between home and work, and public and private life are challenged, patriarchal controls are exposed and undermined in new fashion because the consumerist culture of capitalism also undermines traditional masculinist privilege. So although global capital, as such, is no friend to women and girls, it unsettles existing gender relations in ways that it cannot simply control. In this sense global capital is tremendously contradictory: it promises freedom and riches to the very people it exploits and degrades while also putting this contradictoriness in view. (Eisenstein 2004, 192)

Having had a period of time to digest the empirical research, scholars need to turn more of their attention to theorizing the gendered nature of knowledge work and its relationship to the wider society. This chapter has offered a perspective on how to initiate this process by concentrating on two sets of relationships. The first addresses connections between approaches that start from political economy and those that take off from feminism. The two differ in their emphasis on social class and class power (which is the focus of political economy) and on gender and the power of patriarchy (which predominates in feminist theory). The second set addresses the connections between work and home, concentrating on how scholars have sought to understand the complexities that characterize this relationship, and how workers and their organizations have attempted to addresses problems arising from the relationship.

The chapter has considered four themes, each of them influenced by political economic and feminist perspectives and each of them dealing with responses to the dialectic of work and home. Our goal was to demonstrate the importance of reconciling political economy and feminist theory, and the need to rethink theories of knowledge work and of postindustrialism from a gendered perspective. There are many ways to accomplish this task. We chose to pursue the relationship between work and home, and although we concentrated mainly on North America, we did so with an understanding that this relationship is increasingly open to question in all parts of the world.

The first perspective pointed to the contribution of feminist theory and political economy in uncovering women's labor in the home and accounting for its value. It suggested that, given the continuing dominance of mainstream or neoclassical economics, a great deal of work needs to be done to expand our understanding of household labor so that it can be effectively valued. Furthermore, the perspective demonstrated how technology has been used to draw women and men into a culture of consumption that promises time-saving but delivers a massive restructuring, and at times a draining, of their time. As a result, people spend more time buying and tending household technology, and more time perfecting their skills in what become multiple areas of expected expertise, such as home and interior design, cooking, and cleaning, as well as computer, media, and telecommunications management. They also spend more time as audiences whose unpaid attention is sold to advertisers who fund the commercial media system. Accounting for this labor would lead to a deeper comprehension of how knowledge work operates and to a sharpening of our critical skills about the meaning of home and the place of technology within it.

Moving from unpaid to waged labor, the second perspective concentrated on how paid labor penetrates the home. Industrialism was born in the movement of production from home to factory, but there has always been movement back and forth between workplace and household. Material production, especially in textiles and other trades associated with women, has often returned to the home. This is still the case in both the richer and poorer nations of the world. But with the rise of telework, knowledge labor has made its move back to the home. So have new forms of material labor, such as the production of computer components and other electronics at the base of the knowledge industry value chain. Wage labor penetrates the home to take advantage of cheap labor, to offload capital expenditures to the home, and to provide an incentive to retain workers. There are good political economic reasons for this development, but it is also centrally the case that wage labor in the home is gendered over a wide range of workers, including the immigrant women who work out of their homes for Silicon Valley tech firms in order to care for a family and avoid the U.S. Citizenship and Immigration Service, the secretaries and clerical workers who telework because they cannot afford day care or because they are primary caregivers for their children and elderly parents, and the growing number of women who telework because bosses require it.

The third perspective drew on autonomous theory to examine the ways technology challenges the ability of capitalism to contain either the home or the workplace within the orbit of the marketplace. The growth and expansion of immaterial labor, largely resulting from the application of digital technologies, blurs the distinction between waged and voluntary labor, between work and play, and between home and work. For the autonomist theorist, this creates the potential to liberate people from the oppression of work, and to free women from the double oppression of waged and unwaged labor. Some will find unattractive the scent of utopianism that emanates from this approach, but it certainly raises important questions about genuine difficulties that capitalism faces in enforcing market rules, such as fighting the twin and interrelated practices of music downloading and video "piracy." It also suggests pathways for rethinking the relationship between gender and technology. This may not amount to a new "Cyborg Manifesto," but it should at least lead to a more complex view than a simple political economy of technology can provide.

The final perspective examined how workers' organizations are responding to the shifting relationship between work and home in the world of knowledge workers. It explored different forms this resistance takes, including the value of convergence among labor unions in the knowledge sector, and the creation of worker associations that may be more valuable in a changed world of gendered information work. These forms of worker response are explored in greater detail in upcoming chapters, which address the historical roots of trade union convergence in the "one big union" movement, the value of trade union convergence in the contemporary world, its limitations, and the growth of worker associations and social movements which are providing new ways for knowledge workers of the world to unite.

Chapter Four

Convergence, Solidarity, and Labor Power: The Dream of One Big Union

When the Canadian Broadcasting Corporation applied in 2000 to consolidate its English-language bargaining units, the largest CBC local quickly came up with a nickname for the initiative: "one big union" at the CBC. When the phrase appeared on statements, newsletters, and the website of the Canadian Media Guild (CMG), it was almost always in quotation marks, presumably to convey a wink and a nudge of irony. Clearly the guild—a local of the Communications Workers of America (CWA) through its Newspaper Guild sector—was nothing like the Industrial Workers of the World (IWW), the revolutionary syndicalist organization of the early twentieth century that marched under the banner of uniting workers everywhere to overthrow capitalism. And it was nothing like western Canada's One Big Union, which the federal government blamed for the Winnipeg General Strike of 1919.

But in stressing the idea that the bargaining-unit consolidation would result in "one big union," the guild was evoking a vision: unity would lead to labor power, however that unity might be defined. In this particular instance, it was defined fairly narrowly. The CBC wanted to roll three bargaining units, each represented by a different union, into one. This would complete a decade-long process of consolidating multiple bargaining units—and in the process, eliminating multiple unions—at the national public broadcaster. The Canadian Media Guild fully supported the idea. According to its president, Lise Lareau, "For many years the CMG has taken the position that one single union is best for all CBC employees. It gives the union increased clout, and given the pace of change in the industry, many of the old jurisdictional lines are becoming obsolete" (Lareau 2001). But although the particular case was narrow, the ideas underpinning the drive toward one big union were not. For unions like the CWA or the Communications, Energy and Paperworkers (CEP) union of

Canada, the idea of bringing together workers or their trade unions across the converging communication and information sector has become a central preoccupation. These unions see it as a way of getting ahead of developments in the workplace, not just keeping up with them, and as a way of building, preserving, defending, or extending labor power. But while the contemporary drive toward labor union convergence is a product of this historical moment, its roots are deep, stretching back into the nineteenth century.

This chapter traces some of these roots, following two main branches that have met and diverged time and again. The first deals with the idea of creating labor power though craft solidarity; the second with the idea of creating labor power through mass or class solidarity. It is important to recognize at the outset, however, that the tendencies toward organizing by craft and organizing by class are not mutually exclusive, or even just opposite sides of the same coin. Rather, they represent a recognition that there are many routes to labor power, and many ideas about what labor power means, or whose interests it should serve, or how. In broad strokes, craft unions have tended to be exclusionary, hierarchical, and committed to an economic agenda of preserving and promoting the work their members do. Class-based workers groups, on the other hand, tend to be inclusive, decentralized, and committed to political, social, and economic agendas ranging from reform to revolution. Craft unions, past and present, share a great deal with class-based workers organizations—not least the fact that both groups see solidarity, size, and control in the workplace as critical to success. There have been significant convergences as well as conflicts in goals, ambitions, tactics, and membership between these two types of organizations. The pull of craft solidarity has at times worked against the push toward mass unions, and at other times has encouraged it. Similarly, the appeal of a mass movement has at times pushed craft unions toward broadening their own horizons, and at other times has left them to pull back into their traditional shop-floor stronghold. But at heart, both groups share a similar purpose: tilting the balance of power from capital to labor, or toward those who do the work and away from those who profit from it.

The chapter begins with an examination of the oldest and most powerful craft union in the communication sector in North America: the International Typographical Union (ITU), whose power was based on control of newspaper production. It was also, ironically, the original "one big union" in communications, at one point representing the full range of editorial and production workers in the nineteenth-century newspaper. By the turn of the twentieth century, however, the union had fractured into a half-dozen highly specialized unions, each controlling one step in the production process. This division of both the work and the jurisdiction proved to be remarkably stable, enduring through much of the twentieth century. Ultimately, though, it contributed to the ITU's inability to withstand the waves of technological innovation that swept in on the rising tide of the information society. The chapter then takes up some of the early North American efforts at building broadly class-based unions, touching on the Knights of Labor, the IWW, and the One Big Union. The first was largely reformist. The second and third were revolutionary, though each represented a unique ideological mix. All three, however, believed that if workers could unite, workers and the working class everywhere would gain. These groups contained internal contradictions that contributed to their downfalls, but their most potent enemies were the forces of capital and government, which sought to rid labor of its most radical elements. The chapter concludes with a discussion of the lessons contemporary trade unions might draw from the experiences of both craft and mass solidarity movements, as they seek ways of organizing in a globalizing economy.

The Call of Craft: The Case of the ITU

Craft unions have been central to the struggle over work in Canada and the United States, and also to the development of national and international labor unions and federations. As Fones-Wolf and Fones-Wolf (1983) note, autonomous craft workers—skilled craftsmen who controlled conditions on the shop floor, including apprenticeships and shift assignments—were the backbone of the labor movement in the nineteenth century. According to Stone, "Craft unions were complex systems for transmitting human capital between generations while restricting access to the craft and protecting its economic value" (2004, 19). Newcomers learned the trade through a system of apprenticeships, graduating to journeyman status and, eventually, to the rank of master craftsman. The training covered not just technique but attitude as well. According to Voss, the result was that workers were initiated into a "moral universe that united both masters and journeymen" (1993, 24). Linked to this was a conception of citizenship, in both an inclusive and exclusive sense of the word. Artisans in the United States believed that the small craftsman's shop was the embodiment of republican values and the training ground for participation in the American republic, and that those outside of it had not been well trained for participation in the new and growing nation. Craft unions played a contingent and contradictory role in the process of working-class formation in the nineteenth-century United States. "Craft workers were at once radical and conservative; craft organization was simultaneously a nucleus for further working-class organization and an obstacle to it" (Voss 1993, 7).

The ITU embodied all these contradictions. Printers were among "the status elite" of workers (Lipset, Trow, and Coleman 1956, 27) and what Hobsbawm

(1964, 272) would call the "aristocracy of labour"—the stratum of the working class that was better paid, better treated, and generally regarded as more respectable, and more politically moderate, than the mass of the proletariat. The job of printing required considerable technical skills, from composing type by hand, one letter at a time, to hyphenating and justifying columns of type, to making the printing plates, setting up, inking and operating the printing press, and, in the case of book printing, binding the finished product. It required literacy at a time when many crafts did not, and when most other workers were illiterate. Printers were what contemporary scholars would call knowledge workers. They manufactured not just goods, but ideas and debate. For some, printing was an entrée into the middle class, a chance to become editors, publishers, entrepreneurs, or politicians.

Printers organizations appeared in North America in the late eighteenth century. In the early nineteenth century, stable locals formed in most large cities in the United States and in the Canadian town of York, which would later become Toronto. Independent locals of printers joined together in 1850 to form a national typographical workers union. By 1860 this included thirty-four locals. The union became the International Typographical Union in 1869 with the addition of Canadian locals (Lipset, Trow, and Coleman 1956, 18; Zerker 1982, 4–6). In the late nineteenth century, it was the only craft union representing publishing workers in Canada and the United States, and it had the potential to be a genuine industrial union long before the term had gained popular circulation. Instead, it focused on deepening its hold on the mechanical production of type. ITU membership peaked in the late 1940s at roughly one hundred thousand. This declined to ninety thousand in 1969, to forty-three thousand in 1983 (Scott 1987, 124) and to less than thirty thousand in 1990 (Grey 1990, 2). In the late 1980s, after vicious internal battles over a series of failed merger attempts with the Newspaper Guild, the International Brotherhood of Teamsters, and the Graphic Communications International Union, the ITU became the Printing, Publishing and Media Workers Sector of the CWA. It now plays a role in the operation of the larger union, continues to serve as bargaining agent for members, and administers a large and wealthy pension plan. But these days, the sector newsletter carries more obituaries than reports of organizing victories.

In part because they are the oldest trade in the field, printers have received more scholarly attention than any other group of unionized workers in the communication sector. Lipset, Trow, and Coleman (1956) studied the ITU's unique two-party political structure and Walsh (1985) focused on the union's lengthy and ultimately unsuccessful merger negotiations with the Newspaper Guild. However, the bulk of scholarly work on printers' unions has concentrated on the relationships among technology, skill, and labor power (see, for example, Barnett 1926; Kealey 1986; Kelber and Schlessinger 1967; McKercher 2002; Russial 1989; Sleigh 1998; Zerker 1982; Zimbalist 1979). Much of this work draws, in some cases critically, on the deskilling thesis posited by Braverman (1974). This is the idea that capital uses machinery, scientific management, bureaucratic rules, and any other means at hand to break up the labor process and shift it away from a work system controlled to a large degree by workers themselves. Despite worker and union resistance, the labor process is reconstituted under management control, stripping workers of their skills along the way. As management strengthens its hold over the conception of tasks, workers are left simply to carry them out. Ultimately, the goal of this deskilling is to supplant living workers with "dead labor"—machines that embody the skills once belonging to workers. Other scholars have argued that new technology can result, simultaneously or sequentially, in reskilling and upskilling as well as deskilling, and that negotiation and resistance can have an enormous impact on the implementation of technological change. In the case of the ITU, the contest over skill offers a compelling narrative arc, from the introduction of the linotype to the launch of the online newspaper. Equally compelling is the narrative of accommodation and resistance. In other words, both deskilling and reskilling occurred, but neither happened simply or without resistance. One of the key reasons was that, even before the earliest days of printing unionism, printers saw the work as their own, not as belonging to their masters or to the people who hired them. They insisted on controlling the conditions of their work, a control that was formalized once they formed a national union. For example, print-shop foremen had to be printers themselves and members of the union, a situation that gave the union, rather than the employer, control over hiring and work assignments. Printers set, operated, and enforced their own apprenticeship system, which meant that they controlled the supply of labor. The printers' union certified a member's credentials and issued "traveling cards" that allowed an individual to move from town to town, or across the Canada-U.S. border, in search of work. These forms of control over who was trained, who was certified, and who was hired helped build solidarity among printers and loyalty to the union and the craft. Printers also had ways of disciplining workmates who wouldn't follow the rules, ranging from ostracism on the shop floor, to expulsion from the union, to providing other local unions with the names of "rats" and "scabs" (Lipset, Trow, and Coleman 1956, 30–31).

Within the craft of printing, however, a distinct hierarchy evolved during the nineteenth century, and this was embodied in their union's structure and social relations. As print shops grew, especially in newspapers, the work became more specialized. Some printers worked with engravings and illustrations; others with the stereotype boxes that were used to make curved printing plates to fit on the new rotary presses; still others set up, inked, and ran the presses. Compositors, who set the type by hand, dominated the early ITU in terms of both numbers and power, and the vast majority of these worked in newspaper publishing. Foremen came from the ranks of compositors, including foremen in the pressroom. Indeed, some compositors saw the other specialists as little more than machine tenders (Baker 1974). In the 1880s, the ITU extended its organizing activities to journalists. Here, its main interest was self-protection. The union wanted to prevent journalists, some of whom knew the technical skills of printing, from operating as scabs in ITU strikes.

In the late nineteenth century, the ITU's control over printing and publishing was challenged by a revolutionary new technology, the linotype machine, which could set type mechanically four or five times faster than a compositor working by hand. This machine revolutionized newspaper production in a stunningly short period. Between 1888 and 1904, pretty well the entire newspaper industry converted to machine-set type (Barnett 1926). Publishers were attracted to the machines for a number of reasons. Because they were so fast, publishers anticipated major savings on per-unit costs. Mark Twain once claimed that a composition machine he invested in "could work like six men and do everything but drink, swear and go out on strike" (quoted in Kelber and Schlesinger 1967, 3). Publishers also hoped to employ lower-paid workers, such as less skilled men, women, or even children, to operate the machines, rather than unionized printers. The linotype, therefore, had the potential to displace the majority of printers then employed in U.S. newspapers—Baker puts the potential displacement at up to thirty-six thousand workers (1974, 27)—and perhaps even to break the ITU itself.

But that didn't happen. Some displacement occurred, mainly among older printers and those who had trouble mastering the machine. But the arrival of less expensive type in societies where literacy was on the rise helped stimulate the demand for printed material. Newspapers grew, and the vigorous newspaper competition at the turn of the century resulted in the demand for more printers, rather than fewer. But in addition, the ITU set out and consistently pursued a strategy that would preserve its control over printing. Rather than fighting the introduction of the mechanically produced type, the ITU decided to control it. A resolution adopted at the annual meeting of 1888 called for the recognition of linotypes and recommended that all locals seek "to secure their operation by union men on a scale of compensation equal to that of paid hand compositors" (Barnett 1926, 9). Over the next few years, the union called for moving to an hourly rate of pay rather than a piece rate, and for a shorter workday. Both were designed to protect jobs and wages. It also experimented with changes in apprenticeship rules. Some locals bought or rented machines so that their members could practice. The union's leaders offered no direct opposition to the machinery, and would not tolerate any from the locals.

According to Kealey, a combination of craft customs and vigorous trade unionism were the keys to the printers' success in retaining their control over the production of type while "cleverly acceding to a process of reskilling and demonstrating their willingness to work the new machines" (1986, 93). At the same time, however, the union was becoming vastly more centralized and bureaucratized—"a union that increasingly defied rank-and-file control" (1986, 91). Centralization was not unique to the ITU, however. In the 1870s and 1880s, as printers came into contact with other unions and associations through connections in the Knights of Labor, they discovered that several other unions were better organized than they were. Zerker writes: "It was disconcerting to printers to see themselves falling behind in union reform and benevolent programs, since they had always regarded themselves as pioneers in the labor movement" (1982, 104). As a result, the local printing unions drifted toward accepting greater central authority. Starting in the mid-1880s, the union hired a chief organizer, put its elected leaders on salary, and set up a central strike fund, which could be distributed to a local only after the international executive approved of the strike. The international also started to play a larger role in contract negotiations, eventually putting its own representatives in place to supervise and direct local bargaining with employers. In effect, the international superseded the locals in local collective bargaining.

Within the union, fracture lines that had been present since the 1850s were widening along two axes. One was the split between newspaper work and other types of printing, such as book and commercial or job printing. Just three years after the National Typographical Union was founded, New York book and job printers, including typographers and press operators, complained of being treated as second-class citizens. They pointed out that 90 percent of the New York local's members were newspaper workers and contended that those workers did not take into account the needs of book and job printers. They asked for permission to set up their own local, but were rebuffed (Baker 1974, 60–61). A second and more critical split was the division between the work of composing type and the work of operating the press. Baker explains that the compositors felt superior because theirs was the parent craft, and the one that required the highest standards of literacy and manual and technical skill. In the larger shops, composing room workers were not trained to run the presses. They tended to dismiss pressroom work—and pressroom workers—as less important. In some locals, the pressmen felt no more welcome than women. (For a discussion of North American and British printers' attitudes toward women, see Burr 1993; Cockburn 1983; and Duffy 2000.) Baker, drawing from a 1912 history of the ITU, concludes that the attitude of the ITU's leaders was that if operating a power press was all a man knew about printing, he "would scarcely be considered a printer" (1974, 62). The pressmen believed that the skills required to do their work were just as significant. In addition, while the hand compositors felt threatened by machinery, the pressmen did not, having worked with powered machinery since the 1830s. Like the book and job printers, the pressmen were a minority in the union, accounting for less than one-sixth of the New York local in the 1850s (Baker 1974, 61), and felt unwelcome in their own union. Press operators elsewhere began to complain to the union that they were not regarded as full-fledged printers. A letter sent to the union's 1856 annual convention sought a national ruling that would treat all printers, regardless of whether they set type or ran the press, as equals. This was rejected. But at its next convention, the union ruled that pressmen were indeed printers and it urged locals to encourage pressmen to become members.

Eventually, the pressmen convinced the ITU to issue separate charters for pressmen's locals in areas where there were enough workers to warrant them. But the union ruled that the pressmen needed the approval of the existing local in the area to get a charter, which meant, in effect, that the compositors decided the pressmen's fate. By 1882, pressmen held nine of the ninety-five ITU charters. The next year, the pressmen proposed that the name of the union be changed to the International Typographical and Pressmen's Union of North America, arguing that the growth of pressmen's locals to date, and the potential of "increasing tenfold," justified the change. This request went nowhere. Two years later, the ITU offered a concession: the job of second vice president would go to a practical pressman. This spurred a wave of organizing—the number of pressmen's locals rose to twenty-seven within a year—but it also provoked discussion among the pressmen about whether they should remain in the ITU. In 1889, thirteen pressmen's locals decided to secede and set up their own union. The unhappy bookbinders had also decided go their own way, seceding in 1892 (Baker 1974, 70–74). Over the next two decades, the stereotypers and electrotypers, the type founders, and the photoengravers broke away as well (Lipset, Trow, and Coleman 1956, 20). The ITU's interest in journalists, meanwhile, waxed and waned. In 1896, it created a special vice presidency for editorial workers, but the job lasted less than ten years. Between the 1880s and the 1920s, it established fifty-nine editorial workers' locals in the United States and Canada, but only a handful survived. In 1923, the ITU surrendered jurisdiction over journalists to the American Federation of Labor (Jacobson 1960, 8–10). The last trade to leave the ITU was the mailers, in the 1940s.

The ITU of the twentieth century was, therefore, a vastly different union than the ITU of the nineteenth century. Its claim to being the centerpiece of printing craft unionism had been largely overthrown. Instead of "owning" the entire range of intellective and technical skills required to put ideas on paper, its focus was narrower. Membership was concentrated in the newspaper, with a minority in commercial and job printing. Within the newspaper, ITU members focused on the mechanical composition of pages. Instead of being one big union in printing and publishing, the ITU was one of many unions, each representing a particular craft within the business, and a different stage in the production process. However, its control over the composition of type was as strong as ever. The ITU ruled the composing room, described by one publisher as "a big funnel through which everything must flow to get into the next edition" (London Free Press 1966). The metaphor is an apt one, suggesting that the quickest way to bring everything to a halt was to block that funnel. When the ITU went on strike—and it was more prone to do so than many other unions (Lipset, Trow, and Coleman 1956, 22–23)—it could shut down newspapers in a flash. In the midst of the Great Depression, labor historians Perlman and Taft (1935, 51) recognized the ITU as having "the most complete control over job conditions of any union in the world." Twenty years later, Lipset, Trow, and Coleman (1956, 24) called it "one of the most powerful unions for its size in the nation." Its hold on work was so secure that the union was even able to get around the 1947 U.S. Taft-Hartley Act, which outlawed closed shops.

It is difficult to imagine, in this digital age, the extent of control the ITU exerted over the newspaper. Consider this: the union was so powerful that for close to a century, the first lesson any reporter or editor visiting a newspaper's composing room learned was never to touch live type. "The new man who does so will be quickly put in his place" (Woods 1963, 19). He might be mildly rebuked, shouted at, or warned that touching live type again could provoke a walkout. Intimidating fellow workers was one thing; intimidating the employer was another. The ITU did both successfully, and over a long period of time. In the 1890s, some material, mainly national advertisements, began arriving at newspaper offices ready made, as cardboard matrices that could be turned into printing plates. The union insisted that everything published in an ITU newspaper had to be set in type in-house. Therefore, newspapers that wanted to use those ads had to have their own compositors set the ads. Because the matrices supplied by the advertisers were used in the printing, the type set by the newspaper's own compositors was immediately destroyed. "This has meant that an advertisement used in 500 publications is reset and destroyed 500 times" (Lipset, Trow, and Coleman 1956, 23). This so-called "reproduction right" was a flagrantly wasteful work practice that amounted to union featherbedding, and everybody in the newspaper business knew it, from the publisher on down. But the ITU fought to protect it, often successfully, well into the 1970s.

In the late 1940s and early 1950s, the ITU became aware that it was about to be hit by a second wave of technological change. As early as 1946, a vice president's column in the union's official magazine warned that photocomposition, known as "cold type" because it used an image of a letter rather than an impression of a letter to put ideas to paper, was on the horizon. The column argued that the ITU's "immediate concern" should be to establish jurisdiction over the machines that produced this type (quoted in Kelber and Schlesinger 1967, 32). Union members interviewed by Lipset and his colleagues in the 1950s were acutely aware of the threats posed by new mechanical devices. One of these was the teletypesetter, which produced perforated tape that could be run through a linotype machine. This raised the prospect of producing type without the need for a typographer. Teletypesetters, phototypesetting, and offset printing began appearing across the continent in the early 1960s and, as in the nineteenth century, the pace of the change was stunning. An estimated one thousand phototypesetting machines performed 2 percent of all typesetting in the United States in the early 1960s; by 1973, fifteen thousand machines accounted for 35 percent of typesetting (Zimbalist 1979, 109). The switch to teletypesetters and cold type was matched, and at times led, by computerization in other parts of the newspaper, including in the newsroom, where editors sitting at computer terminals began to set columns of type. These rolled off the phototypesetters as strips of type, already hyphenated and justified. All the compositor had to do was trim and paste up column-width strips onto page-width forms. These new technologies displaced substantial numbers of workers. A study conducted by three Ontario newspaper unions estimated that the move to paste-up production reduced the size of newspaper composing rooms by 50 to 60 percent (Pane 1994, 27). Almost all of this displacement occurred through attrition, early retirements, and buyouts rather than through direct layoffs. But taken together, new technology offered the printing industry "potential independence from the skills of the linotype operator" (Zimbalist 1979, 107) and from the need for the ITU.

The union responded to these challenges in much the same way it had responded to the linotype. It agreed that publishers should be able to introduce the new technology, but the union would do whatever was necessary to retain control over the work. The union called its strategy "controlled automation." Kelber and Schlesinger (1967), in a study focusing on the New York ITU local known as Big Six, write that in its fullest form, controlled automation had a number of features. First, the union concentrated on winning complete jurisdiction over every new machine or process that impinged on or replaced existing methods of getting type ready for the press. As with the linotype, jurisdiction was seen as critical to maintaining ITU members' authority over the setting of type and the composition of pages. The union also took steps to train its members to operate the new equipment, opening a new training center in 1956. As new technologies appeared, the center expanded and grew. According to Kelber and Schlesinger, "Emphasis of the ITU on training was based on the assumption that the skilled printing craftsman who could master the new equipment and processes had nothing to fear, even in an era of rapid technological change. The source of union strength, it was believed, was in the ability of ITU locals to supply printing employers with the most qualified labor available anywhere" (1967, 71). Providing the most qualified workers meant, of course, that the union could continue to control the supply of workers. Big Six also insisted that a publisher who wanted to introduce new technology had to win the prior approval of the union. Kelber and Schlesinger note that the local was the only one of the ten unions working at the New York newspapers in the early 1960s to gain this veto. Finally, the union demanded a financial share in any benefits that the new technology might create, and it refused to water down economic demands in exchange for job security (1967, 267–69).

Controlled automation appeared clear on paper, but the strategy was far from certain to succeed, even in New York. One of the results was a 114-day newspaper strike that shut down all the dailies in New York in 1962–1963. Media coverage tends to frame strikes in ways that reflect an antilabor, proconsumer bias (Martin 2004). Tracy (2006) found just such a pattern applied to coverage of the 1962–1963 New York newspaper strike. The leader of the ITU local was presented as greedy, self-aggrandizing, and at times irrational, while the chief "victims" of the strike were the citizens of New York, who were robbed of their newspapers. Settling the strike did not resolve the situation for either the printers or the newspaper readers. Another year of strife broke out in 1965 and ended with the amalgamation or closing of roughly half the city's dailies. Sleigh (1998), focusing on the *New York Daily News*, describes almost continuous clashes between labor and management of the city's dailies during the 1960s, 1970s, and early 1980s. Most centered on some aspect of technological change.

One of the worst failures of the controlled automation strategy occurred in Toronto, where a strike/lockout that began in 1964 ended up throwing the union out of the newspaper business in that city. Zerker (1982) shows that the ITU's international contracts committee insisted on upping the ante during contract negotiations with the three Toronto publishers in 1962 and 1963. The most troubling issue the union raised in the talks was jurisdiction over computers. At the time the union put it on the table, none of the Toronto publishers had ordered a multipurpose computer. Once the issue was raised, the local union and the publishers tried to reach agreement on the matter. But each time the local's negotiators thought they had a deal, the international union's contracts committee rejected it. Eventually the publishers decided they didn't want to have to deal with the ITU any longer. They unilaterally imposed new work rules, knowing this would drive the ITU to strike. What happened next offered a lesson in why it was much more difficult for the ITU of the 1960s to control automation than it had been for the ITU of the 1890s. For one thing, the publishers kept publishing. Through a combination of hiring new staff (including, though the publishers denied it, a handful of professional strikebreakers), contracting out to commercial print shops, and hands-on work by supervisors, the newspapers continued to hit the streets. Indeed, the publisher of the *Toronto Star*, Joseph Atkinson, claimed that the ability of the papers to publish "without appreciably adding staff" demonstrated just how much streamlining could be done in the composing room (quoted in Zerker 1982, 290).

To the disappointment of the striking workers, all other unions but the mailers crossed the ITU picket lines. The crossing of another union's picket lines is always a difficult subject, and this was especially so in the newspaper business where one union, the ITU, was more powerful than the others. The Newspaper Guild, which represented journalists and white-collar workers, generally tried to avoid crossing picket lines set up by other unions. The ITU, on the other hand, had an internal law that said it would live up to its own contracts, regardless of what other unions did, which meant that the ITU would cross another union's picket line and go to work. There had, however, been exceptions. In 1950, the ITU and other unions refused to cross a guild picket line at the *New York World-Telegram* and *Sun*, which ensured the guild a victory. In 1965, the ITU Baltimore local originally agreed not to cross the guild picket line, but was ordered to do so by the international, thereby crushing a guild strike. In Toronto, "the arms-length, tenuous rapport that characterized the compositors' association with the other printing craft unions" would exact "a terrible price" (Zerker 1982, 280). The ITU's insistence on claiming jurisdiction over any new technology that impinged on printing had put it into conflict with the other unions and was guaranteed to create more battles—immediately in some newspapers, and eventually in all of them. Shortly after the Toronto strike began, the publishers offered the printers lifetime job guarantees in exchange for allowing publishers to install new equipment and processes. The printers would also have to give up their so-called reproduction right and control over the appointment of foremen. The union turned the offer down, but in doing so lost a lot of the public's sympathy. The strikers published a weekly strike newspaper in hopes of rallying support to their cause. Not only did it fail to do so, but according to the strike committee it drew criticism "almost universally from other unions whose cooperation we now seek" (quoted in Zerker 1982, 286). The strike, which began in 1964, dragged on for years. The ITU finally admitted defeat and called it to an end in late 1971.

Where controlled automation seemed to succeed in New York, at least for a time, it failed utterly in Toronto. But with hindsight, it is clear that it was not a workable strategy in the first place. One set of reasons had to do with the technology; the other with the unions. It was one thing to successfully claim jurisdiction over the linotype. Linotypes were single-purpose machines. All they did, and all they ever could do even in their most modern incarnations, was set type, and they could only do so in the hands of skilled workers who understood the art and craft of typography. It was quite another matter to assert jurisdiction over computers. Digital technology could be used for multiple purposes, for tasks as different as calculating the payroll, tracking classified ad sales, and adjusting ink levels on the presses. Most significantly for the ITU, computers could set type. Editors who had never been trained in typography could, at the touch of a few buttons, set headlines that were the correct size to fit the space, and hyphenate and justify columns of type. Zimbalist writes that with the arrival of computers in the 1960s, "the eventual extinction of all craft vestiges became a certainty" (1979, 108).

The newspaper entered a period of rapid technological change, each stage of which used digital technology to simplify, streamline, reduce, or eliminate various stages in the production process. Technological change did not mean simply replacing one production process with another; rather, it meant collapsing previously separate stages into each other and moving as much production as possible to the news desk. This round of technological change happened incrementally and continuously, over a number of years (McKercher 2002). The depth and diversity of the new technology meant it was impossible for the ITU to "control" this round of automation. It is interesting to speculate on whether things would have been different had the ITU stayed together at the turn of the twentieth century. As a genuine industrial union, representing all unionized workers in the newspaper business, the ITU might have stood a chance of exerting more control over the pace and scope of technological change—or at least of controlling the division of the remaining work among the surviving production workers. But in workplaces where it was but one of several unions, any such control was impossible. In addition, the ITU's strategy of fighting with the other unions over jurisdiction proved to be damaging. It led to years of wrangling, in newspaper after newspaper across North America, between unions that should have been allies, over a dwindling share of work.

The ITU's insistence on seeing itself as the centerpiece of organized labor in printing and publishing did little to help. In the early 1960s, ITU president Elmer Brown raised the idea of creating—or to be more accurate, re-creating—one big union in printing and publishing. He frequently referred to the idea as an invitation to the other crafts to "come on home" (Walsh 1985, quoted in

McKercher 2002). This phrase was a reminder of the ITU's historic position as the parent of the printing unions. But it cast the other unions in the role of runaways or disobedient offspring, who were now being asked to return to the parental home and, at least possibly, to parental control. Brown also insisted that any merger would have to be on a basis that would not "dilute" the ITU (quoted in Zerker 1982, 322). Other unions chafed at what they saw as arrogance among ITU leaders. Sleigh quotes a New York journalist of the 1970s who no doubt spoke for many when he expressed his resentment of Big Six leader Bert Powers's view of the communication business as "just another factory job for the printers" (1998, 93). Zerker, writing in the early 1980s, was struck by the fact that ITU printers were still caught up "in that strange fantasy about the 'mighty' ITU and its entitlement to dominance over all other allied trades" (1982, 322). The Toronto printers believed in "myths that told of the might, loyalty, generosity, and wisdom of the ITU" and their "almost fanatical loyalty" to the union was a critical feature of the tragic 1964 strike (1982, 279).

And yet it is important to recognize that this self-perception had a lot to do with the union's ability to build and maintain the loyalty of generations of members, and to preserve and even encourage craft solidarity, even in a rapidly changing industry. Kelber and Schlesinger (1967) note that controlled automation included a refusal to water down economic demands in exchange for job security. That changed in the 1970s, when the union began accepting offers of lifetime job guarantees from publishers in New York, Washington, Ottawa, Vancouver, and elsewhere. This made sense for the publishers. In the long term, new technology offered major savings in labor costs, though everyone knew it would not happen overnight. The hope was that by promising lifetime jobs, the publishers would gain workplace peace and a cushion of trained staff to manage the transition. In the long run, of course, retirements and buyouts would finish the job of replacing workers with equipment. In 1974, Big Six finally agreed to give the publishers a free hand to install new technology, and to give up its "reproduction right" for national advertisements, a source of irritation for decades. The price to the publishers was an eleven-year contract, wage increases, and lifetime job guarantees for any composing-room worker displaced by technology. Though controlled automation was finished, a generation of printers won the guarantee of work for as long as they wanted.

The ITU stands as an example of a union that successfully built and maintained control over craft, and in the process garnered the loyalty, trust, and support of its members. This enabled it to put up an impressive fight against powerful publishers armed with new technologies, a combination that would have crushed a less powerful or less self-confident union. In the end, the ITU was gutted by new technology. But that came as no surprise to anyone: industry analysts writing as far back as the early 1960s had predicted this. What was surprising, though, was how long the printers were able to assert their claim to the work, and how successful they were in influencing the scope and pace of change. As the ITU's struggles over technological change demonstrate, craft control can indeed be a route to labor power. But craft control means more than control over the tools of production. It also encompasses control over the supply of workers, through training and apprenticeship programs. In addition, in the ITU's case it also meant control over the way those workers were assigned and deployed, through its insistence that foremen be union members. Even more significant was the ITU's ability to instill loyalty, solidarity, and pride in its members. Lipset, Trow, and Coleman, in a survey of job satisfaction among ITU members, found that 96 percent liked being a printer, and that 73 percent liked it "very much." By contrast, a survey of American male workers found that although close to 90 percent of skilled trades workers were satisfied with their work, only 32 percent felt "very satisfied" (1956, 137). This sense of pride in the craft is one of the important legacies of craft unions like the ITU.

The Call of Class: Dreaming of "One Big Union"

Although unions based on craft have been central to the struggle over work in Canada and the United States, and to the development of labor federations at the national and international levels, the other route to labor solidarity—the various incarnations of the dream of one big union—continue to spark the imagination of contemporary scholars, trade unionists, and activists. The Knights of Labor, the key player in a remarkable if short-lived flourishing of labor movement culture in the 1880s and 1890s, has received some scholarly attention, much of it critical (see Kealey and Palmer 1987). The Industrial Workers of the World (IWW), a radical syndicalist organization known as the Wobblies, has inspired a renewed interest in labor among students and has even made its way into popular culture in recent years (Buhle 2005). Evidence of this can be seen in Kellerman's (2007) forty-page bibliography of works on the IWW, which includes scholarly books, biographies, novels, poetry, song lyrics, and so on. The occasional use of the IWW's "one big union" slogan by communication-sector unions, ranging from the ITU and Newspaper Guild in the early 1960s to the CBC workers in 2000, reflects an enduring fascination with the idea of working-class solidarity as a route to labor power, if not with the IWW itself. In addition, proponents of social movement unionism, an increasingly significant feature of the contemporary labor movement in North America, recognize that they are pursuing many of the same goals as the broad-based labor movements of the past. As we will show in chapter 6, today's social movement unions have a political, economic, social, and cultural agenda that includes organizing the unorganized, making labor rights synonymous with human rights, and tearing down the walls that separate the workplace from the community. All of these were features, to a greater or lesser extent, of earlier attempts to create "one big union."

The first successful attempt at creating large-scale, class-based unionism in North America began in 1869, the same year that the National Typographical Union became the International Typographical Union and more than fifteen years before the creation of the American Federation of Labor, the national federation of craft unions that would dominate organized labor in the twentieth century. The Noble and Holy Order of the Knights of Labor was launched in Philadelphia by a handful of members of the dissolved Garment Cutters Association who shared a vision of a new kind of organization. As Weir puts it, "The founders dreamed of moving beyond the limits of 'bread-and-butter' unionism by creating an oppositional culture based on the brotherhood of all toilers" (1996, 9). Even then, solidarity across craft lines was not a new idea: the National Labor Union, which sought to unite workers from all trades into a single organization, had been established three years earlier, though by the time the Knights were founded it was on the verge of collapse. But what was original about the fledgling Knights was that the organization combined class-based unionism, bringing together skilled and unskilled workers within and across industries, with the idea of a fraternal brotherhood, based along the lines of the Masons. The Knights believed that secrecy and ritual, two hallmarks of the Masonic movement, could become powerful tools to create membership loyalty and organizational stability, as well as a means of hiding activities from employers.

The early Knights were exclusively male, white, and skilled. That changed as the organization moved north and west, especially into the mines of Pennsylvania. In 1878, at its first Grand Assembly, the group adopted a constitution that, among other things, called for the abolition of child labor in factories, workshops, and mines, and equal pay for equal work by men and women. The preamble to the Knights' constitution, which operated as a statement of principles, committed the group "to secure to the toilers a proper share of the wealth that they create; more of the leisure that rightfully belongs to them; more societary advantages; more of the benefits, privileges and emoluments of the world; in a word, all those rights and privileges necessary to make them capable of enjoying, appreciating, defending and perpetuating the blessings of good government" (Knights of Labor 1878). The preamble painted a picture of an organization committed to equality, social justice, opportunity for workers to improve their economic lot, and substantial reforms to the relationship between labor and capital. Key ideas included the establishment of cooperatives, the protection of public land from exploitation by the railroads and speculators, the elimination of laws that favored capital over labor, and the enactment of laws to protect worker health and safety. The constitution also called for arbitration rather than strikes, "whenever and wherever employers and employe[e]s are willing to meet on equitable grounds." The Knights' agenda was reformist rather than revolutionary, and it was almost synonymous with the populist movement prominent in many rural areas in the final years of the nineteenth century. This emphasis on the social world set the Knights apart from the craft unions, which focused on the workplace and tended to have little interest, as organizations, in broader political questions.

In the 1880s, the Knights began to move away from the early stance of secrecy and the numbers began to grow in earnest. Assemblies formed on both sides of the Canada-U.S. border, and on both coasts of the continent. One reason for growth was that membership was open to almost all, including other trade unionists: only bankers, lawyers, gamblers, speculators, and liquor tradesmen were barred (Weir 1996, 11). Individual Knights joined local assemblies, either mixed assemblies (open to a broad range of workers) or trade assemblies (limited to specific craft categories). Local assemblies were coordinated by district assemblies, and the supreme authority was the national assembly. The Knights sought to attract the largely disenfranchised mass of unskilled workers. Women and African Americans, two groups who were, in general, spurned by traditional craft unions were welcome to join. Local assemblies were not required to organize in all skills, which meant that each defined its own collective identity. Some local assemblies were limited to African Americans or women; others were mixed. Most immigrants were welcome, but not those from China. Fourteen local assemblies established in British Columbia in the mid-1880s specifically excluded Chinese laborers. In a statement to an 1885 Royal Commission, the Knights denounced these workers as "low, degraded and servile" and as "willing tools whereby grasping and tyrannical employers grind down all labor to the lowest living point" (quoted in Palmer 1992, 124).

Membership grew rapidly, from 9,300 in 1879 to 50,000 less than five years later (Weir 1996, 10). Many of the new local assemblies were organized in places with a strong craft union tradition. Conell and Voss (1990) found that the greater the number of organized craft workers in a given location, the greater the strength of the corresponding local assembly of Knights. This suggests both a labor culture that encouraged organizing in general, and the likelihood that resources from craft workers helped the Knights. Many craft workers who organized under the auspices of the Knights had dual affiliation, belonging to a craft union as well as to the Knights. Membership in the Knights hit 74,000 in 1884 and more than 111,000 in 1885, the year it launched a successful strike against Jay Gould's South West Railway conglomerate. According to Weir, "Gould's defeat caused a stampede to the Knights that can be likened to the faithful embracing a savior" (1996, 12). Within a year, membership jumped to 730,000. Local assemblies formed so quickly that the general executive board declared a moratorium against chartering new assemblies. But this did little to deter enthusiastic would-be Knights, "who met without benefit of charters, or the burden of dues-paying" (Weir 1996, 12). Probably a million individuals called themselves Knights. This included thousands of Canadian workers, mainly in Ontario, Quebec, and British Columbia. Adding up peak memberships from local assemblies, Palmer estimates that that the group organized at least 21,800 workers in Ontario alone, but adds: "A figure double, perhaps triple, this tally is likely a more accurate reflection of the number of workers that passed through the order" (1992, 123).

The Knights soon paid the price for rapid growth. Many new members did not know the organization's history, or even some of its central tenets, such as the commitment to arbitration rather than strikes. Instead, the successful strike against Gould led many to see strikes as the best and quickest tool to achieve labor reform. Leaders of the Knights warned that strikes were hard to win and should be used as a last resort. Many local assemblies ignored this advice, however, and the result was a wave of strikes, capped by a call for a general strike over the issue of the eight-hour workday. The Haymarket incident in Chicago—in which a bomb wounded more than seventy policemen during a demonstration by strikers, prompting the police to open fire on the crowd—triggered a massive backlash against organized labor. Meanwhile, the Knights themselves were plagued by intense ideological debates and factionalism, especially in New York. This led to "a loosened sense of solidarity with the rest of the labor movement that spilled over into jurisdictional fights with trade unions" (Weir 1996, 13) and with the American Federation of Labor (AFL), which was then a new organization, founded in 1886 (Kaufman 2001). The number of Knights plummeted, to 120,000 in 1890 and down to 80,000 in 1893 (Voss 1993, 79). In its later days, it returned to a ritual secrecy like that of its early years. A few local assemblies survived well into the twentieth century, but the group eventually faded away.

In assessing the reasons for the Knights' collapse, Voss concludes that the key factor was not the weakness of the organization but the relative strength of American capital. It countered the threat of an organized working class with effective organization of its own, mobilizing powerful employers' associations. Weir draws a similar conclusion: "It was its success against capital that did in the Knights, not internal conflicts" (1996, 327). After 1885, as membership grew faster than the Knights could manage, capital's resolve to crush the movement deepened and employers closed ranks. Palmer (1992) argues that at their height in the 1880s, the Knights were part of a genuine working-class "movement culture" that grew out of both a growing recognition of class status and the social upheavals brought about by rapid industrialization and the consolidation of capital. This movement never had the time to develop a strategic response to capital's orchestrated effort to break the movement. Nonetheless, Voss (1993) contends that the defeat of the Knights demoralized those who championed radical reform and classwide organization. It also empowered those who promoted pragmatic politics and sectional labor unions, especially the American Federation of Labor. This goes a long way toward explaining the "weakness and conservativism" of the American labor movement from the turn of the century through the mid-1930s (1993, 249).

Palmer offers a different interpretation. Many trade union pioneers were both Knights and craft unionists, he notes. "To stress the craft unions' 'victory' over the Knights of Labor and to argue that a committed and intellectually aware stratum of North American labor leaders consciously chose the narrow path of trade unionism over the politically oriented, reformist purpose of unifying all workers is to mistake historical outcome for strategic direction" (1992, 149). Certainly the leaders of the AFL and its Canadian counterpart, the Trades and Labour Congress, and the leaders of many of the unions that belonged to those federations, tended to be more conservative than the Knights. But there was a great deal of diversity within the labor movement. The political views of rank-and-file craft workers ranged from traditional republicanism to anarchism, from parliamentary socialism to Marxism. In the ITU, for example, a split between radicals and conservatives resulted in the creation of a two-party political system within the union that was one of its unique and most enduring features. On the job, meanwhile, rank-and-file union members behaved in ways "that can hardly be described as conservative" (Kimeldorf 1999, 4), pursuing aggressive job action at the point of production to win the day. At the same time, labor radicals outside the AFL were busy organizing unskilled workers, especially miners. The independent Western Federation of Miners advocated industrial unionism and launched or supported efforts to form an industrial union of railway workers and two industrial union federations (Bercuson 1978, 39). In Canada, British labor radicals were as influential as their American counterparts, spreading the message of revolution and class conflict. In both countries, a range of radical political organizations helped shape and popularize the principles of revolutionary labor syndicalism—the notion that class struggle was inevitable, and that workers needed to join together in one great union, which would create a revolution and serve as the basis for postrevolutionary society.

Revolutionary syndicalism's clearest expression in North America appeared in the form of the Industrial Workers of the World (IWW), which was founded in Chicago in 1905 and moved into Canada a year later. At its high point in 1917, the IWW counted more than one hundred thousand members (LeFevre 2005), or roughly one-tenth the membership of the Knights of Labor at that organization's peak. This was nonetheless a remarkable number, given the revolutionary nature of the organization. Where the Knights were reformist, the IWW was not: its goal was the general strike that would topple capitalism. Its conception of "one big union" was a unified, revolutionary working class. But until the revolution could begin, workers needed a better deal, including higher wages and better working conditions, especially in the dangerous and dirty railway and mining camps. The combination of revolutionary goals and trade union ideals presented the IWW with a fundamental conflict. Negotiating with an employer on behalf of its members or other members of the working class would in effect recognize capital, and therefore legitimize capitalism. Holding true to its revolutionary principles, the IWW rejected measures that would have ensured its growth as a trade union. It refused to sign contracts, because signing a contract meant accepting the employer's right to own the factory. It rejected the idea of the dues checkoff system, or the automatic collection of dues by an employer, on the grounds that it stripped workers of their autonomy. The IWW refused to create a long-term strike fund, in part because it feared that the fund could be seized by the state but also because it felt that money encouraged conservatism, which would get in the way of class struggle. It did not provide unemployment or sick benefits, on the grounds that this would make it more difficult for workers to develop class consciousness. All of this meant that the IWW lacked a stable financial base and professional leaders, and it could not guarantee long-term membership (Leier 1989). In addition, IWW leaders fought with each other about the direction the organization should take. The result was "a constant state of turmoil at the IWW's upper levels" (LeFevre 2005).

The IWW's propensity to strike—some joked that IWW stood for "I Won't Work" (Laut 1912/2003)—infuriated not just the businesses affected by strikes but other industrialists and factory owners as well. Canada's entry into the First World War in 1914 opened the door to the repression of IWW leaders as radicals and enemy aliens. When the United States joined the war in 1917, repression of the IWW in that country "started with a vengeance" (Leier 1989, 118). Hundreds of members were arrested on charges of, among other things, conspiracy to obstruct the war. Four mass trials sent more than 160 IWW members to jail. In 1918, Canada outlawed the IWW and other left-wing groups through an Order in Council, in effect a government edict that, unlike a statute, is not subject to parliamentary debate. More labor radicals were arrested and deported from the United States and Canada in 1919. Remarkably, the IWW survived, and has even experienced periodic spurts of membership growth. But it has never regained its hold on the popular imagination of the working class.

Radical trade unionism reappeared in Canada briefly after the war in the form of the One Big Union, founded in 1919. Though it attracted many of the same workers as the IWW (Peterson 1983), One Big Union represented a different ideological mix, described by Leier (1989, 119) as "the fusion of industrial unionism and parliamentary socialism that many in the Socialist Party of Canada had long advocated." This union was the product of western radicals, who had less experience than easterners with craft unions and believed that the road to labor power required the amalgamation of crafts into fewer, larger industrial unions. These radicals initially tried to reform the AFL-linked Trades and Labour Congress and, when that failed, they created their own organization. The group, which at its peak included fifty thousand members, quickly became a focal point for "anti-Bolshevik hysteria" (Bercuson 1978). In May 1919, the Winnipeg trades council called a general strike, which lasted until late June and set off a wave of sympathy strikes in the four western provinces. The government concluded that the One Big Union had been created to organize and coordinate these revolutionary activities, though in fact three of the five leaders of the Winnipeg General Strike were what Bercuson calls "faithful craft unionists" (1978, 95), a fact that offers vivid evidence of just how diverse were political views among craft union members. The collapse of these strikes hurt the One Big Union and helped expose conflicts within its leadership. The organization was further weakened by a craft union counterattack, strongly supported by business and government (1978, 261). After the crafts froze the One Big Union out of negotiations, its influence declined rapidly. Remarkably, the One Big Union endured until the mid-1950s, but as "a living fossil" in the labor movement (1978, 247).

There are a number of common themes in the stories of these various attempts at creating one big union. The Knights, IWW, and One Big Union all suffered internal dissension, as leaders disagreed over goals, strategies, tactics, politics, and agendas. They all ran into difficulty providing adequate services to members. All had problems dealing with the rival vision of syndicalism offered by the craft unions, which sought to create a better society by reforming the workplace rather than the other way around. Employers refused to recognize these larger organizations, encouraged conflicts between the craft unions and the mass unions, and at times joined forces to combat them. The stance of the governments, meanwhile, ranged from grudging tolerance to outright suppression. Nonetheless, these organizations had a major impact on skilled and nonskilled workers, and on the larger society. Palmer (1992) concludes that the greatest achievement of the Knights of Labor was to bring the "labor question" to importance. That is the enduring legacy of the IWW as well.

Industrial Unionism in a World of Convergence

The industrial unions that characterize the contemporary labor movement represent an effort to bring together the ideas of solidarity based on a mass movement and solidarity based on craft. Industrial unionism rose in the 1930s, springing up alongside a set of social, economic, cultural, and political reforms that grew out of the New Deal. These reforms included legislation that recognized workers' rights to join trade unions, formalized the process for doing so, and set the ground rules for collective bargaining. Industrial unionism at its simplest is predicated on the idea that all workers within a particular industry—steel, for example, or rubber—should have the opportunity to belong to the same union. Industrial unionism sought to organize all workers, not just craft workers, and—like the Knights of Labor—it put special emphasis on organizing women and minorities. As a social and economic force, industrial unionism grew rapidly, quickly outstripping the AFL's ability to contain it. The Committee for Industrial Organization, created within the AFL in 1935, became the Congress of Industrial Organizations (CIO) just three years later and served as a rival labor federation to the AFL until the mid-1950s when the two groups reunited. The president of the ITU was the first secretary of the CIO, though he served as an individual, not as an official representative of his union. Nonetheless, the ITU was one of eight unions that were ejected from the AFL for continuing to push industrial unionism, remaining outside the craft federation until 1944. In a climate that recognized organized labor's right to represent workers, the new CIO doubled its membership from 800,000 in 1936 to nearly 1.4 million in 1940. Meanwhile, the AFL grew at almost the same rate, from 2.6 million members to 4.2 million (Kimeldorf 1999, 163). The enthusiasm for organizing was equally strong in Canada, even though the nation lacked the kind of protection offered by the 1935 U.S. Wagner Act until the 1940s. Membership in unions belonging to the CIO's counterpart doubled between 1939 and 1943. Many of the large manufacturing unions that would dominate the merged AFL-CIO and Canadian Labour Congress after the mid-1950s emerged out of the industrial unionism of the 1930s. Denning (1996) makes a link between the rise of industrial unionism and a "laboring of culture" in North America during the era from the Great Depression to the 1950s. He notes that many of the labor unions representing cultural workers—including teachers, newspaper reporters, motion picture actors, and radio stars—originated in the 1930s. Labor issues moved to the forefront of the cultural struggles, and the labor of cultural production in particular gained a new prominence. This helped to consolidate labor as a political and economic force.

Industrial unionism was, however, never a fully realized process. In the communication sector, as in many others, multiple unions continued to represent groups of workers, and rival unions fought over jurisdiction. Until the 1970s, for example, newspaper workers were represented by the white-collar Newspaper Guild, founded in the 1930s, and by up to nine different blue-collar unions, including the ITU and many of its offshoots. Though the number of unions has been reduced, multiple jurisdictions continue to be common in the newspaper business today. U.S. telephone workers were originally represented by company-dominated employee associations until the Wagner Act outlawed such organizations in 1935. Thereafter, the Bell companies helped the workers create 184 "new" associations. A critical mass of these joined to create the National Federation of Telephone Workers in 1938. It became a genuinely national union, the Communications Workers of America, in 1947 (Miller 1998). But it was by no means the only union interested in representing telephone workers. At the end of the Second World War, the United Electrical Workers sought to disband the telephone workers' federation as a company union. The CIO established its own telephone workers' organizing committee. And the International Brotherhood of Electrical Workers successfully organized the Western Electric manufacturing operation in New Jersey. The bulk of Canadian telephone workers were represented by the Canadian branch of the CWA, which separated from the United States union amicably in the 1970s and became the Communications Workers of Canada—and ultimately, the "communications" part of the CEP union. In British Columbia, however, telephone workers founded their own union, the Telecommunications Workers Union. Meanwhile, the Canadian Auto Workers union represents telecommunications workers at Manitoba's MTS Allstream.

In the performing arts, the Screen Actors Guild, the Directors Guild of America, the Writers Guild of America, and the American Federation of Radio Artists (AFRA) appeared in the 1930s, joining older unions like Actors' Equity as bargaining agents for stage, screen, and radio artists. Some broadcast journalists belonged to the radio performers' union, AFRA, which later became AFTRA when jurisdiction was extended to television; others would eventually find their way into the Newspaper Guild. In Canada, the picture was even more complex. U.S.-based international unions dominated the trade union movement in Canada until the 1960s. Though their influence has declined with the creation of Canadian unions, these unions are still a major force. As a result, unionized workers in the cultural industries in Canada are represented by a mix of Canadian unions, including the CEP, the Canadian Auto Workers, and the Alliance of Canadian Cinema, Television and Radio Artists (ACTRA), and by "internationals" like the CWA and the International Alliance of Theatrical Stage Employees.

Industrial unionism arose as an effort to fill a vacuum in the manufacturing industries, especially in mass production. Prior to the mid-1930s, the industries

Brody characterizes as "the mass production core"—iron and steel, nonferrous metals, rubber, electrical products, chemicals, petroleum, and food processing—were impervious to unions. Over the next decade, industrial unions appeared in almost all of them (Brody 1993, 82). Organizing unions by industry made sense in an economy that was based on manufacturing industries. It is considerably more problematic, however, in an information society—one that is characterized by the production, distribution, and exchange of intangible commodities rather than manufactured goods. In the communication and information sector, the lines that once separated media products from each other have been blurred. So have the lines that once separated media industries. Newspapers are parts of conglomerates with telecommunications companies and broadcast or cable television networks; movie studios are mixed in with record companies and television networks. Drawing clear lines between industries, and hence between what should delineate the scope and jurisdiction of industrial unions, is increasingly difficult. And in some cases, perhaps, even pointless.

Some scholars studying ways to reinvigorate the labor movement from the current state of decline described in chapter 1 have looked to the energetic intellectual and social mix that characterized the industrial unionism of the 1930s, not just to the ways the unions divided the terrain (see for example Brody 1993; Krupat 1999; Turner and Hurd 2001). This was a time of phenomenal growth in trade unions, but more significantly it was a time when the labor movement was part of a social movement that sought broad-ranging reforms to improve the lives of working people, their families, and their communities. Many of the industrial unions that were created in the 1930s, however, would later become institutionalized, professionalized, and bureaucratic, operating primarily as official bargaining agents for members. In this chapter, we have looked to earlier forms of labor organization that called for many of the same kinds of reforms as industrial unionism. Our examination of the rise and fall of the ITU, the Knights of Labor, the IWW, and the One Big Union, has focused on exploring how to build labor solidarity. The question for the contemporary labor movement is whether these organizations offer any lessons that could be applied today in the converged knowledge and communication arena.

From craft unionism, for example, comes the idea that craft identity matters to workers, and can be a tool for building stronger unions. The remarkable success of the ITU was its ability to make workers identify with their union—to see the union, not the employer, as the institution that would provide them with a good living, and therefore to see the union as worthy of an individual's support, trust, and loyalty. The ITU taught newcomers their craft and certified their skills. But more important was its ability to instill in workers an identity with the craft, a sense of its importance in the printing and publishing business, and a belief that the union would take care of its own, from the time a worker joined the union to the time it came to pay for the member's funeral. However, craft solidarity can also be destructive. The ITU newspaper compositors tried the patience not just of their employers but of their fellow workers, through decades of fighting to claim jurisdiction over the tools of the trade. Nonetheless, it is clear that when workers see their union as protector and promoter of the work they do, rather than simply as a bargaining agent, they are more likely to see the union as central to their lives. This encourages solidarity and, perhaps, trade union democracy, since engaged trade union members are more likely to be interested in participating actively in the union.

One of the key challenges facing converged unions in the communication sector is making members feel that their union knows their concerns, and can speak effectively on their behalf. The CWA's approach to this has been to create separate "sectors" for previously independent unions that merge with the CWA. This allows them to preserve their unique culture, leadership styles, and member services while at the same time interacting with other information workers and taking advantage of the services of the larger union. Canada's CEP union integrates units that move over from other unions more fully into the power structure of the union, but at the same time has initiatives to recognize their unique needs. The union has developed a code of ethics for journalists, for example, and it keeps an up-to-date, online database on media ownership in Canada. Both are of interest chiefly to media workers, but are available to all members of the union. At times, the CEP draws on its status as an all-Canadian union to promote solidarity and loyalty among its members. It has used this particularly effectively in encouraging Canadian locals of U.S.-based internationals—from the National Association of Broadcast Employees and Technicians (NABET), the Newspaper Guild, the ITU, and, most recently, the Graphic Communications International Union—to secede from their parent unions and join the all-Canadian CEP instead. In short, unions that recognize that culture counts, including craft culture and national identity, stand a better chance of engaging and mobilizing their members.

The massive expansion and sudden collapse of the Knights of Labor serve as a warning that growth has to be managed carefully. The Knights obtained members at a pace that far outstripped their ability to integrate them into the organization. Growth also outstripped the leaders' ability to understand the needs of some of the organization's new members, especially craft workers with concerns that were specific to the work processes in their plants. Ironically, the larger the organization grew, the weaker it became. For converged unions in the communication and information sector, the lesson to be drawn is that growth does not automatically translate into labor power. It has to be managed carefully and accompanied by a commitment to serve the needs of new members, especially when those new members work in areas that have been, until now, peripheral to the union's original mandate. This can become an issue both in trade union mergers and in new organizing. As we show in the next chapter, the "one big union" created at the CBC worked mainly because the CWA local that emerged as the sole bargaining agent worked hard at integrating the members of the other unions, especially the technicians who had previously belonged to the CEP. Both unions had waged a hard-fought campaign to become the sole union for the English-language broadcast workers. But once the representation election was over, both sides had to work out ways, not just to tolerate each other, but to draw on each other's strengths and build a cohesive union. Unless new members feel as welcome in, and as central to, the mandate of their new union, their commitment to its agenda may be questionable.

A third theme that emerges from the experience of both the craft unions and the early efforts at "one big union" is the strength and agility of the opposition: mainly the employers, but the state as well. This serves as a reminder that labor organizations not only have to keep track of what their employers are doing, but also need to anticipate what they are likely to do next. They need to look outward, at what is happening in other parts of the communication sector, and in other parts of the world. As national economies become integrated internationally, and increasingly reliant on digital technology, work cannot be neatly divided into national industries. In a globalizing world, capital does not stop at the border. Nor should the effort to build labor power. The challenge for trade unions is to counter the agility of the opposition with an agility of their own, based on building bridges across industries and across national borders.

Chapter Five

Labor Convergence in the Information Economy

In chapter 2 we set out ways to understand the growth of convergence in communication technologies and among communication companies, and how these developments have led to an increasingly integrated electronic information services arena, dominated by a handful of global corporate conglomerates. Companies have used the power generated by mergers and concentration to eliminate some jobs, contract out others, cut salaries and benefits, and deskill many of the jobs that remain (Craypo 1997; Krishnan and Park 2002; McKercher 2002; O'Shaughnessy and Flanagan 1998. For a gendered dimension on mergers and organizational change, see Collins 2005. For the role of The North American Free Trade Agreement [NAFTA] in facilitating this process, see the Economic Policy Institute [EPI] 2006; Shoch 2001, esp. 281). One bright spot for workers is the growth of convergence in the labor movement, and especially in the communication and information sectors.

This chapter examines the process of convergence within traditional trade unions, including the degree to which different communication and knowledge industry unions have adopted it and the consequences for unions and their members. It begins by sketching the rough outlines of labor convergence in the communication and information sector in the United States and Canada. This shows up in various forms, including mergers among national unions, bargaining-unit consolidation within individual unions, and the activities of national or international labor federations. It then takes up three case studies in trade union convergence. The first, focusing on the Communications Workers of America (CWA), is, broadly speaking, a success story. The second, an examination of the failed merger efforts between two key Hollywood unions, offers lessons on the negative fallout that may result from the failure of convergence. The third, a look at the Telecommunications Workers Union's struggles against the Canadian telecommunications giant Telus, describes mixed results arising from incomplete convergence. The chapter concludes that labor union convergence is difficult to achieve, full of contradictions, and offers no guarantees of success at the bargaining table. Nonetheless, it can indeed be a powerful force for working people, and it has the potential to allow unions to "bite back" at companies that have used technological and industry convergence to chew up and spit out their unionized workforce.

Convergence and Divergence in North American Trade Unions

Labor convergence takes many forms, depending on the nature of the organization and the local conditions it is facing. But perhaps the most high-profile form is the merger of communication unions across sectors within one country. In the United States, several media unions, including the International Typographical Union (ITU), the Newspaper Guild, and the National Association of Broadcast Employees and Technicians (NABET) have joined the Communications Workers of America (CWA), which began life as a telephone workers union and now has seven hundred thousand members. Calling itself "the union for the information age"—the phrase is the banner on the union's website, www.cwa-union.org—the CWA has expanded across the communication industry so that it now represents workers employed in telecommunications, broadcasting, cable TV, newspaper and wire service journalism, publishing, and electronics. The union has also expanded beyond the media and telecommunications industries to include general manufacturing, as well as airline customer service, government service, health care, and higher education. As we shall see, the move into the service sector set the groundwork for a new and expanded version of convergence, one that incorporates the electronic information services arena. Among the companies employing CWA members are AT&T, the Regional Bell telephone companies, Sprint, the NBC and ABC television networks, the Canadian Broadcasting Corporation (CBC), and major newspapers such as the *New York Times*, the *Wall Street Journal*, and the *Washington Post*. CWA employers also now include the full range of customer services and call-center firms central to the airlines and insurance industries, and to the marketing divisions of consumer products and services companies throughout the economy.

In Canada, the motivation for convergence has been tempered at times by a desire for distinctly national unions. Many of the country's major private sector unions are the result of decisions to secede from international parent unions, headquartered in the United States. For these new Canadian unions, therefore, the first step toward convergence was divergence or separation from the American unions that once ran their affairs. Once independent, these unions have grown in their own particular ways, sometimes diverging significantly from the pattern set by the original U.S. parent union, and sometimes mirroring it. Canada's largest private-sector union, for example, is the 265,000-member Canadian Auto Workers (CAW), which broke away from the United Auto Workers in 1985. Since then, it has more than doubled in size by undertaking thirty-seven mergers that added 150,000 members. In addition to its core in the automotive sector, the CAW now represents workers in road, water, rail, and air transportation; the aerospace industry; media and telecommunications; fisheries, mining and smelting, and shipbuilding; hospitality and gaming; retail and wholesale trade; general manufacturing; health care; and colleges and universities (CAW 2008). The UAW has also broadened its focus, as its current name reflects. It is now the International Union, United Automobile, Aerospace and Agricultural Implement Workers of America (UAW). The Communications Workers of Canada, which until the 1970s was the Canadian branch of the Communications Workers of America, joined with two other unions in the early 1990s to create the Communications, Energy and Paperworkers Union of Canada (CEP). It has pursued a convergence strategy that is similar to that of the CWA, though with its own unique features. For example, the CEP merged with Canadian NABET and with many of the Canadian units from the ITU and the Newspaper Guild. When the Graphic Communications International Union decided in 2005 to merge with the International Brotherhood of Teamsters, much of its Canadian membership chose to join the CEP instead. CEP has the potential to represent communication workers of every sort, at every stage in the production process, and at every level of skill. CEP members, for example, make newsprint, work in the growing energy industry that powers the information society, and keep the telecommunications system working. They write and edit the news, and put it on the air, on paper, and on the Internet. They are also employed as graphic artists, hotel workers, computer programmers, truck drivers, and nurses (Swift 2003).

The Telecommunications Workers Union (TWU), which originally represented telephone workers in British Columbia, has taken a different and less robust approach to convergence. It has incorporated workers throughout the telecommunications sector, including the addition of wireless workers to its traditional wireline workforce, and it has expanded into several additional provinces. Specifically, it was able to extend its jurisdiction over telecommunications workers in the province of Alberta because the Canadian labor regulatory body, the Canada Industrial Relations Board (CIRB), determined that workers experiencing technological and industry convergence were best represented by one merged union, even if that meant crossing provincial borders. The union also moved further east, into Ontario and Quebec, following rulings by the CIRB supporting labor convergence in telecommunications (CIRB

1999, 2000a, 2000b, 2004a). But it has chosen not to extend its organizing beyond the telecommunications industry.

The motivations behind labor union convergence, and the results unions expect from it, vary from union to union. Some may see their actions as ways of protecting members from the growing power of the companies that employ them. Others may view labor convergence as a means of providing members with protection from governments that have taken a probusiness position—a frequent occurrence in a period of neo-liberalism—or as a way of strengthening labor's voice in the policy arena. Still others may see union convergence as an attempt to take advantage of synergies brought about by growing convergence in their work and their workplaces. Since they represent workers who are increasingly involved in producing for a converging arena of electronic information services or knowledge, these unions see improved opportunities for organizing and for bargaining.

The CEP has recognized that convergence also provides unions with the resources and the political capabilities to be more involved in a wide array of social and political activities, including those in conflict with the narrow but hard-felt interests of one or more of its constituents. Moreover, Swift (2003) cites the CEP as a case in point of a converged communication union that has been more deeply involved in major policy issues since it expanded over the converged information industries. For example, the union has produced research and lobbied forcefully to limit media concentration in Canada, in spite of the fact that one of its major employers, Bell Canada Enterprises (BCE), is also the largest media conglomerate in the country, with substantial control over the local and long-distance wired and wireless telephone networks and Internet service, as well as stakes in the leading national daily newspaper and the major private television network in Canada. To support its campaign to advance professional journalism in the face of the media concentration threat, in 2007 the union published one of the largest-ever studies of the state of Canadian journalism and of the views and working conditions of journalists (CEP 2007).

Even though the CEP represents energy workers, it stands fully behind the Kyoto Accord to limit the expansion of greenhouse gases and has fought against all government and corporate attempts to scuttle it. The strong support for climate-change policies coming from the national office appears to have filtered down to the membership, including those working in the energy industry itself. The union represents workers at one of Canada's foremost oil projects, which is involved with extracting the product from tar sands in Alberta, an operation that is one of the leading producers of carbon dioxide emissions in Canada. Nevertheless, as one tar sands worker put it, "The national office gives us a sense of direction. It's good to have the national telling us it's important to support the Kyoto Accord even though we're in the oil industry" (Swift 2003, 20).

The union has also used the resources of convergence to work for its members on policy issues that are less contentious within the union. As chapter 3 noted, the union is a leader in fighting for pay equity and, in 2006, it won its fourteen-year pay equity battle against Bell Canada (Freeze 2006). The company agreed to pay up to $100 million to almost five thousand women who worked for the company during the 1990s. Convergence gave the union the resources necessary to take on such a long, hard, and expensive battle against the leading unit in Canada's largest conglomerate. Furthermore, the CEP has also fought hard against lifting restrictions on foreign ownership of Canadian telecommunications and electronic media. The union has maintained a close tie to the Canadian Centre for Policy Alternatives. Julie White's (2005) report for the Centre on "Losing Canadian Culture: The Danger of Foreign Ownership of Telecom" provides a good example of how convergence offers an opportunity to support labor-relevant research at a progressive think tank. The CEP has also been in the forefront of lobbying to maintain public telecommunications in the province of Saskatchewan and public hydroelectric power in Ontario. It was able to stand up for its paper workers against the powerful wood-products company Abitibi because convergence permitted the CEP to draw from the strike funds of its energy and communication industry members. It also has the resources to create a Quebec Solidarity Fund that permitted it to invest in declining Quebec paper mills and keep them from closing. The union has been extensively involved in the antiglobalization movement and in supporting unionization in Mexico and throughout Latin America, with the help of the CEP Humanity Fund. Additionally, research conducted by Kiss and Mosco (2005) on what unions are doing about surveillance in the workplace has demonstrated that knowledge-worker unions, especially convergent unions like the CEP, provide the best privacy protections for workers in their collective agreements.

As with the corporations that employ them, there is also a mythic aspect to the notion of labor union convergence, tapping into an enduring fascination with the idea of one big union. In chapter 4 we described how the mass-based movements of the nineteenth and early twentieth centuries saw this as more than just a strategy or policy: "one big union" embodied a vision, a dream or grand narrative about uniting industrial workers into a mass social movement that would transform labor power into political, social, and cultural power. Today, the vision has returned. But now it embodies the hope of bringing together workers in one large national or binational organization of knowledge, communication, and information service employees. In essence, converging employers, converging media products, converging jobs, and converging technologies have led workers to unite across the previously divided sectors of the knowledge industry (McKercher 2002). The unions pursuing labor convergence have promoted it to unionized and nonunionized workers alike, as a way of strengthening a union's ability to resist the depredations of employers, and as a means of enabling unions to play a stronger role in social movement organizations that seek to stem the erosion of public service communication and information services.

This labor convergence strategy has not always followed the pattern of unions merging across different media sectors within one nation. It also occurs through attempts to solidify a union's position with a particular employer, often through bargaining-unit consolidation. The term refers to efforts to extend a union's jurisdiction to workers who are either not represented by a union at all, or represented by a different union. One example that sparked some controversy and is central to one of the cases described in this chapter is the absorption of all unionized English-language workers at Canada's national public broadcaster, the CBC, into the Washington, D.C.-based CWA. The process by which this took place and its implications for a crucial labor dispute at the CBC are examined below, but for now it is important to highlight two reasons that convergence at the CBC was significant. First, it marked an important example of binational convergence. Second, notwithstanding the success, one of the enduring controversies about bargaining-unit consolidation at the CBC is that all the workers employed by a national public broadcaster, including on-air as well as technical personnel, now belong to a union headquartered in the capital of another nation. The CBC case provides an example of how the process can work successfully across borders, but it continues to trouble some advocates of national Canadian unions.

Canada's CEP and TWU unions have also engaged in bargaining-unit consolidation. On the west coast, the CEP emerged in 1996 as the sole bargaining agent for all unionized workers at two large Vancouver daily newspapers now owned by Canwest Mediaworks Publications. CEP initially fought alongside the six other unions at the company, then known as Pacific Press, to prevent a bargaining-unit consolidation at the *Vancouver Sun* and *Vancouver Province*. But when the British Columbia labor board ordered a consolidation vote, the CEP unit not only emerged as the sole union at the plant but took on the challenge of becoming the leading media union in the province (McKercher 2002). On Canada's east coast, CEP fought successfully for the right to represent wireless workers as well as conventional telecommunications workers. At about the same time, the TWU fought for succession rights to the staff who worked for the Telus company in Alberta, Ontario, and Quebec following a series of corporate mergers. As we noted above, this came about with the support of Canada's federal labor board (Mosco 2002, 2005; Mosco and McKercher 2006; Shniad 2007).

Another form of trade union convergence takes place when federations of existing unions come together into working coalitions, largely because of technological convergence and common interests. An example is Canada's National Alliance of Communications Unions (NACU), which brings together the CEP, the TWU, and one local of the Canadian Auto Workers into a federation of telephone and telecommunications workers. In the United States, convergence led in 2005 to a federation within a federation when the AFL-CIO, itself the largest trade union federation in the country, created an Industry Coordinating Committee made up of ten constituent unions in the arts, entertainment, media, and telecommunications industries. Again, the recognition of common interests in addressing a converging industry, alongside the need to preserve craft independence, motivated this development. In a similar vein, ten communication unions in Canada have created the Coalition of Canadian Audio-Visual Unions, which is concerned about the erosion of the public service principle in the Canadian media industry, and specifically with the steady demise of Canadian content regulations (Coles 2006).

Recognizing the limitations of national or even binational unions, labor organizations are also converging internationally. If workers around the world can work for a single multinational employer, the most effective strategy, supporters maintain, is for those workers to coordinate and act collectively on an international scale. They can do this by forming worldwide union networks to confront these multinationals. Arguably the best example of this is the Union Network International (UNI), founded from the merger of four global unions in 2000. Based in Nyon, Switzerland, UNI has incorporated 15.5 million workers in 900 unions in 140 countries, covering 12 general economic sectors, predominantly those in the knowledge economy. It operates as a confederation, with each sector maintaining autonomy to choose its own president, vice presidents, and committees and to determine its own policies, priorities, and structures. For example, the media, entertainment, and arts sector deals with such issues as media ownership and concentration, structural evolution of the sector, trade policies, legal standards, technological change, copyright, and censorship. UNI's telecom sector handles liberalization, deregulation, privatization, and mergers and acquisitions—all forces driving the restructuring of global telecommunications markets and policies. Industry-wide coordination on a global scale is seen as an effective way to build real worker power and is the underlying rationale for establishing these sectoral groups. But at the same time, UNI promotes what it calls a "global union agenda" that covers global organizing, global framework agreements, global objectives for fair work and rates of pay, minimum standards for health and safety, universal collective bargaining rights, global solidarity action, global policy development, and global representation (UNI 2005h). UNI has become one of the central nodes in a global network of labor opposition to offshore outsourcing, a business strategy that is significantly eroding labor's power and a major policy issue facing governments around the world. Chapter 7, which addresses global labor organizations and the changing international division of labor, provides a detailed analysis of UNI's activities in the knowledge and communication sector.

Trade union convergence is far from easy to achieve, however. It requires unions, which often see each other as competitors and may have a history of "raiding" each other's members, to set aside their differences and work together. Mergers are particularly difficult to achieve. In a merger, one or both of the unions must voluntarily give up some degree of autonomy, learn how to share power, shed traditions and practices that helped distinguish the union from others, make changes in the way the union operates day to day, and in many cases give up a cherished trade union identity in exchange for another, newer, and less familiar identity. The film and video industry in the United States provides an important example of just how difficult this can be. The Screen Actors Guild (SAG) and the American Federation of Television and Radio Artists (AFTRA) know that one of the keys to mobilizing against the increasingly integrated video and film industries encompassing the television networks and the Hollywood majors is to present a unified front. In the absence of labor unity, employers are better able to play one union off against the other, and to deflect complaints about shooting a television series or a feature film in Toronto or Vancouver, Prague or Budapest, rather than in the United States. The two unions have worked together for many years, and a majority of members agree that the next step should be a full merger. But all attempts have failed, and in the years since the last failed merger attempt, in 2003, relations between the two have become increasingly hostile. In Canada, the attempt to build closer ties among major telecommunications unions through the National Alliance of Communications Unions also has not been especially successful. The two key unions, the CEP and the TWU, have not worked closely together. One reason may be the latter's history of radicalism: in 1981 the union took over and operated the telephone exchanges of Vancouver during a labor action. In addition, the TWU has tended to be skeptical about the value of the convergence projects undertaken by the CEP. For this kind of federation to work, member unions have to be equally committed to pursuing common cause. In this instance, that does not appear to be the case.

Clearly, even among communication and information sector unions that endorse the principle of convergence or see it as an idea that is worth pursuing, there is disagreement about how seriously to pursue it, how to accomplish it, what form it should take, and how successful it can be in strengthening labor's hand. Having set out in broad strokes the outlines of convergence, we now turn our attention to some examples of the difference convergence—or the lack of it—makes in specific industries.

Convergence at Work: The Communications Workers of America

The Communications Workers of America (CWA) provides the best example in North America of a union with a focused convergence strategy. It concentrates on the communications sector but defines it broadly. CWA members range across the spectrum of creative and technical workers and encompass a growing share of the service sector, which the union has committed to drawing into its ranks. By bringing together workers across the various communication industries, the CWA has expanded its resource base and its expertise to deepen and extend its organizing activities.

The union developed this strategy in stages, over a number of years. From its founding until the late 1970s, it was mainly a telephone and telecommunications union. But starting in about 1980, it began to expand beyond the industry, first into government work, then into health care, universities, law enforcement, airlines, and other areas. In 1991, it declared that its goal for the decade was "wall to wall" organizing—in other words, to unionize all workers at a company, not just one group. The results of that strategy continue to be seen today. In April 2007, for example, the union's Newspaper Guild sector enrolled 220 workers who write and edit wire copy for Dow Jones Newswires, thereby completing the unionization of the company. What is particularly interesting in this organizing drive is that the group also includes technical, sales, and support personnel as well as reporters and editors.

In addition to deepening its traditional organizing activities, the union has applied its convergence strategy to branch out into some challenging new areas. Chapter 6 describes in some detail the growth of worker associations in the communication and information technology sector, including the CWA's relationship to a group representing workers at Microsoft, but it is useful to mention this development here because it signals both the strengths and the limitations of a broad national convergence strategy. The computer and information-technology sector has been notoriously difficult to organize, even in the bust periods of its boom-bust cycles. There are a number of reasons for this, including a strong streak of individualism among information-technology workers; unfamiliarity with, or antagonism toward, unions and union culture; a sense of professionalism; and a belief that the work, unlike unionized work, offers an individual the potential to strike it rich. When information-technology workers are drawn to the outer edges of the political spectrum, they tend to explore libertarian ideals rather than the social democratic ones that have historically inspired union organizing (Brophy 2006; Rodino-Colocino

2007). No less significant is the power of companies like Microsoft and IBM to fight off unionization drives, sometimes directly but more often by promoting a corporate culture that emphasizes employability within a profession rather than stable employment in a single company, and that discourages individual workers from doing anything that might threaten their employability. Nevertheless, with the assistance of the CWA, workers at Microsoft and IBM have formed organizations to defend workers' rights and provide services to members. And while neither the Washington Alliance of Technology Workers (WashTech) nor Alliance@IBM has won the right to engage in collective bargaining at their respective companies, they count several successes in the courts and in political lobbying that have justified the CWA investment in their activity. These associations have also been instrumental in their own forms of labor convergence. For example, WashTech organized wireless workers at Cingular. Admittedly, success in the high-tech field has been limited, and the ability of employers to outsource jobs internationally, as Microsoft and Amazon did in response to WashTech organizing efforts, has demonstrated the limits of what amounts to a national convergence strategy. But the ties that the CWA has been able to create in the information-technology sector have the potential to extend its reach, as well as to provide the union with an opportunity to experiment with new forms of worker organization.

Convergence has also enabled the union to expand into the service occupations that depend heavily on communication and information technology. Conceptually and historically, labor convergence at the CWA has meant joining workers across the barriers that separate communication forms: adding broadcast workers to newspaper workers to telephone workers, for example. But now it also means uniting service workers from a number of different industries and sectors, including, of course, the communications industries, but also government, health care, and transportation services as well. All told, the union represents 160,000 customer service workers. Regardless of which industry employs them, they are increasingly doing the same kind of work—talking with customers by phone, and working from scripts that look more and more alike, regardless of the product or service. As CWA research director George Kohl puts it, "What's the difference between being a telephone customer service agent, or an airline customer service agent, or a customer service agent in the newspaper industry? There is no difference—you're all customer service professionals" (Interview with authors 2007). As a result, the CWA has been trying to build what Kohl characterizes as "a craft and programmatic identity" for the union's customer service workers:

> So we've sponsored for the last couple of years a customer service week, which is worldwide through UNI. We're on our third or fourth one, trying to describe what is a positive vision of a customer service model in terms of working con-

> ditions or work rules. There are common elements that overlap and that should be our programmatic force. How do you deal with monitoring? How do you deal with workplace pressures in terms of different systems? (Interview with authors 2007)

Customer service work is particularly significant to the CWA for two reasons. First, the call-service industry has been growing rapidly, and hence offers opportunities for organizing. Second, the international outsourcing of call-center work has also been growing rapidly. The CWA has taken steps, admittedly small ones, to build ties to workers in the major destinations for outsourced work, especially in India. In 2006, it participated in a joint study with Indian labor organizations, notably the New Trade Union Initiative and Jobs with Justice in India, which sought to understand the scope and extent of outsourced service work from the perspective of workers in both the United States and India. The study recognized that workers in the United States and India are, for the first time, linked in the global supply chain of customer service work. Its objective was to identify shared interests and to counter the view that labor unions in the United States see foreign workers as the enemy. The study noted that the organizations participating in the binational study share a "global concern that unionized work should not be outsourced and that outsourced work should not remain non-union." The report drew on a 2004 survey of call-center workers in the industrial nations and carried out fresh surveys of Indian workers to assess the state of the call-center industries in both countries. Building on this research, the participants in the project came up with a series of recommendations for improving working conditions, wages, benefits, and standards in both countries (Centre for Education and Communication et al. 2006).

The CWA's commitment to convergence has allowed it to take a leading role in studies of this type. In recent years, it has also expanded its activity in policy-relevant research, including establishing a close relationship with what is arguably the major progressive think tank in Washington, the Economic Policy Institute. In 2007, the union stepped up its involvement in debates about Internet policy by producing *Speed Matters*, a detailed analysis of issues, problems, and policy alternatives in the U.S. broadband industry (CWA 2007). The report found that even as Americans were paying more for slower service, the United States fell from eleventh to sixteenth place among nations in access to high-speed Internet services from 2002 to 2005. To remedy this shortfall, the report offers an ambitious program to expand connectedness, speed, and quality of service. It would be easy to dismiss the report as self-serving: after all, the CWA and its employers would be among the chief beneficiaries of the massive development project the report recommends. But the depth and quality of the study are difficult to dismiss, and it does not shy away from controversy. Following the general tenor of the progressive media reform movement in the United States, it supports a level playing field and an open Internet for all users:

> Government policies should enable the widest possible dissemination of information from a multiplicity of sources over the Internet as the foundation of a democratic society. Moreover, since the value of any network increases as more content is transmitted to more people over its paths, unrestricted Internet access is good economics—stimulating consumer demand for more bandwidth and reducing supplier costs through economies of scale. Because the greatest threat to an open Internet is bandwidth scarcity and unequal deployment of high-speed networks, building and maintaining universal high-speed networks must be the first priority in protecting and strengthening an open Internet. (CWA 2007, 26)

Interestingly, the union stops short of endorsing the network neutrality policy that has been a centerpiece of the communication reform movement in the United States. The network neutrality standard is based on the idea that if information networks are to maximize human welfare, they ought to be as neutral as possible between various uses or applications, with no schemes for differential pricing, speed, and quality of service. Moreover, government ought to intervene to promote or preserve the neutrality of the network. Instead of joining most other media reform organizations, the CWA report argues that the need for network investment requires special incentives for corporations:

> Since private capital will largely finance the high-capacity network build-out, network providers must be assured a return on their investment. That return comes from value-added services and, in current business models, primarily from video. Policymakers, therefore, should refrain from imposing open-network requirements on network providers' offering of video and other private network services. Such policies would dampen investment in job-creating high-speed networks and other specialized services such as voice telephony, video streaming, medical monitoring that operates more effectively with quality of service and other managed network upgrades. (CWA 2007, 26)

However one comes down on this issue—and there are many different shades to the question—it is clear that the CWA is a significant participant in the debate, choosing to use its resources to intervene on policy matters that affect members across several sectors.

The CWA has shown how convergence can contribute to building the membership, expanding international worker solidarity, and improving the quality and quantity of research and the opportunities for policy intervention. But perhaps the most important benefit is the contribution that convergence can make when workers are forced to, or decide to, withhold their labor power in a dispute with their employer.

In 2005, the CWA faced a case that tested its commitment to convergence and its ability to operate in the international arena when Canada's national public broadcaster, the CBC, locked out its workers. The CBC faces stiff competition from private radio and television broadcasters but, unlike its counterpart in the United States, it plays a leading role in the Canadian broadcasting industry. Although its audience share has not always kept up with the private sector, the CBC is a dominant force in news and public affairs, in both radio and television.

It is not unusual for workers to take a convoluted route to convergence and the CBC is no exception. Jurisdictional issues have dogged the national broadcaster for two decades, with twelve different unions representing workers at the corporation in the 1980s. This often left workers in similar jobs laboring under very different employment rules. In 1993, in response to management's desire to create a more harmonious workplace, the federal government's industrial relations regulatory authority reduced the number of unions representing English-language CBC workers to three: the Canadian Media Guild, a unit of the Newspaper Guild (which was itself in the process of merging with the CWA), would represent editorial workers; the Canadian Union of Public Employees (CUPE) would represent clerical and administrative workers; and the CEP, through its merger with Canadian NABET, would represent technicians.

Shortly after this, the CBC went through a major restructuring triggered by funding cuts from the federal government. Waves of voluntary departures and layoffs washed over the workforce in 1996, 1997, and 1998, mainly among the administrative staff. The remaining members of the CUPE unit asked to join the Canadian Media Guild (CMG) and the guild agreed. The two groups continued to be divided into separate bargaining units, however. In 1999, the guild filed a petition with the federal labor board to merge the administrative and editorial workers into a single bargaining unit. The CBC countered with a proposal to unify all of the bargaining units: CEP as well as the CUPE and CMG units. In making its case for consolidation, the corporation told the labor board this would result in increased efficiency, increased mobility, a common framework of working conditions, increased industrial stability, better coordination of work overlap, common supervisory responsibilities, better operational coordination, a similarity in collective agreement provisions, and the reduction or elimination of jurisdictional disputes arising from the blurring of responsibilities between units. Put simply, it would streamline and facilitate bargaining (CIRB 2003).

In essence, this meant that the CBC asked for trade union convergence at the company. The CMG supported the national broadcaster's petition, agreeing that one bargaining unit would be a better arrangement. As a local of the CWA, the CMG position dovetailed with the convergence strategy and the "wall to wall" mandate of the parent union. Of course, the guild was aware that it would most likely emerge as the bargaining agent: CMG members outnumbered CEP members by about two to one, which meant the guild was almost certain to win a vote to choose which union would become the bargaining agent for all. The CEP, in part for the same reason, objected to the consolidation. It also argued that the bargaining-unit structure put in place in the early 1990s was working. The system wasn't broken, so there was no need to repair it.

In February 2003, the CIRB sided with the CBC and ruled that the bargaining-unit structure was no longer appropriate (CIRB 2003). The CEP asked, unsuccessfully, for a reconsideration of the ruling. In November 2003, ballots went out to eligible employees asking them to choose which union—the binational CMG or the all-Canadian CEP—would represent them all.

The campaign leading up to the vote was bitter and divisive, with the CEP expressing outrage that a "foreign" union should seek to represent workers at Canada's national public broadcaster. This was especially difficult because the CEP itself was the product of a hard-fought effort to break away from American dominance and establish genuine national unions in Canada. Ironically, its members at the CBC were former NABET members, who had left their American counterpart and found their way into the ranks of the CEP. Their NABET counterparts in the United States were now part of the CWA, which meant that the Canadian members would end up rejoining the very union they had left. Meanwhile, the CMG members bristled at the suggestion they were somehow and suddenly "un-Canadian" because their parent union, the U.S.-based Newspaper Guild, had merged with another U.S.-based union, the CWA.

As expected, the guild had the votes: the final tally was 2,452 for the CMG and 1,717 for the CEP. In essence, the CBC and the CWA "ganged up" on the CEP to force a consolidation, with everyone knowing all along that the CEP would be the loser. At around the same time, the CBC's French-language division, Radio-Canada, sought a similar consolidation to simplify bargaining among its unions, but the CIRB turned down the broadcaster's position because it concluded that there had been a history of labor strife at Radio-Canada which might be deepened by all-or-nothing bargaining, and that none of the unions supported Radio-Canada's proposal for union convergence (CIRB 2005).

When the dust settled, CBC management had "won" a single bargaining unit with which to negotiate contracts for its English-language radio, television, and Internet workforce. But in doing so it now had to face a consolidated bargaining unit that was part of one of the largest, and certainly one of the most powerful, private-sector unions in North America. It remained uncertain just what kind of a unit it would be: Would the wounds inflicted on both sides in the representation election leave permanent scars? Or would the two sides put down their weapons and work together?

Almost immediately after the vote, the CMG launched a campaign to integrate the CBC branch from top to bottom. Arnold Amber, Canadian director of the CWA, says it held a conference in Montreal in April 2004 that got people from all sides together to talk about what kind of union it should be, and followed that up with a meeting in Ottawa that began to put a new structure in place. It made sure that CEP people were welcomed onto the unit executives in the cities where the CBC operates, and were encouraged to run for office. It revised the structure of the CBC branch. And because the CBC is by far the largest of the eight branches that constitute the CMG, it added CEP members to the national executive as well. Amber says the integration effort worked well.

> After people get over the loss of their old union identity, and after the other side, the side that wins, is willing to say, "Let's forget the campaign and let's forget the old stuff. So you called me tremendously horrific things during this four-month campaign. I'm going to forget that, and you're going to forget our argument, and let's go on," you'll find that the reason people were in the union is the intrinsic values of unions and their sense of being involved in the union. There were small numbers of people who actually couldn't make that leap, but by and large people could. (Interview with authors 2007)

Because the union was going into negotiations with the CBC, it also revised the structure of its bargaining team, reserving places for former CEP members who had intimate knowledge of the old CEP contract, could speak for the technicians, and could speak to their concerns.

The CBC's goal in the bargaining round that began in 2004 was to fundamentally transform its fifty-five-hundred-member workforce. This transformation had actually begun earlier, with increases in the number of nonpermanent or temporary workers, growing integration of radio and television operations, and experiments in video journalism that drew on individuals who combined the skills (and did the jobs) of both journalists and technicians. These practices fit the corporate convergence model because they permitted greater flexibility in the company's ability to move workers in and out of different units and job categories. CBC management wanted to deepen and extend this transformation by weakening many of the rules protecting workers' professional status, particularly by winning the unlimited right to hire temporary workers for all of its jobs. Management's objectives were not simply rooted in the desire to control its labor force. Rather, they had been driven by a decade or more of budget cuts, combined with growing pressure to do more with less—to deliver the same or a greater level of service, now extending from broadcasting in English, French, and a number of Aboriginal languages, to maintaining a significant presence on the Internet, with fewer resources. The CBC was, of course, not unique: budget cuts and pressures to commercialize public institutions "force" the attack on labor, and the attack on labor makes it easier for governments and arm's-length entities like the CBC to cut spending and promote the overall commercialization of operations.

Bargaining began in May 2004 and although the parties met regularly, numerous issues remained unsettled a year later, including the key issue of a contingent workforce. The CBC applied for conciliation in May 2005. The union took a strike vote in July, which the membership overwhelmingly supported, and on August 15 the CBC locked out its employees. The company's timing of the lockout was strategic. Parliament was in recess and therefore unlikely to intervene until it returned in the fall. In addition, the CBC feared that the workers would go on strike in September, at the start of the new fall television season, so the company calculated that it was preferable to get the dispute over with in August. What the corporation had not foreseen, however, was that the labor convergence it helped to create would "bite back" by unifying a once-divided workforce into one of the most successful converged unions on the continent. With the support of the CWA, which provided close to $10 million in strike benefits, workers could afford to stay out longer. In addition to paying for half of all strike pay and for all of the supplementary health benefits of locked-out workers, the CWA made full use of its international networks to support the CMG. It mobilized an international pressure campaign that directed messages from media workers throughout the world to CBC management and the Canadian federal government. It organized demonstrations at a dozen Canadian embassies, including Washington, D.C., London, and other power centers.

Convergence also promoted unity among the workforce. Where they were once based in different unions headquartered in different countries, which promoted invidious comparisons, all workers now succeeded or failed based on how well they worked together in their struggle against the CBC. While it would still be possible to play journalists against technicians and vice versa, it is demonstrably easier to do so when they are members of different, competing unions. Furthermore, labor convergence made it more difficult for management to operate the system during the lockout. It was one thing for management to take over technicians' jobs when the on-air staff remained on the job, as happened in an earlier CEP dispute, but quite another when the entire workforce was locked out. Consequently, audiences felt that the lockout resulted in a significant erosion of CBC services, particularly when management was forced to turn to BBC broadcasts, with practically no Canadian content, for its news programming.

The deployment of temporary or contract workers is central to a corporation's convergence strategy. Even though the scholarly literature has raised serious questions about the actual organizational benefits of such flexibility for the broadcasting industry, one reason corporations pursue a convergence strategy is to increase their political flexibility in dealing with workers (Ursell 1998). Indeed, once it realized that employees were unified, CBC management offered to protect the jobs of all existing permanent employees and increase the use of nonpermanent labor for new positions only. However, this decision backfired. Workers resisted, arguing that an increase in temporary workers would seriously impair the CBC's ability to fulfill its public service mandate, and would ultimately damage civil society in Canada. Once Canadians realized that the workers were staying out to support the right to full-time jobs at the national broadcaster, even though these workers had been given immunity from more stringent labor practices, public support for the locked out workers grew (Delacourt 2005).

Another characteristic of communication labor that motivates management to support convergence strategies also backfired in this case. Communication workers are knowledge workers—part of the growing "creative class" who produce information and entertainment products—and information-technology workers, especially those who develop software to run converging information networks. For management, one of the great challenges of the information society is how to control creative labor and lower its cost through various forms of labor division, deskilling, antiunion activity, and contracting out (Barley and Kunda 2004; Braverman 1974; Brint 2001). CBC chose to expand nonpermanent contract labor as a major way to strengthen its control and lower costs. But in doing so, the corporation challenged a value dear to creative workers: the right to creative control over their own labor, to retain the flexibility to make use of it in ways they think are most beneficial to fellow citizens. To demonstrate that the commitment to full-time professional labor is an essential quality of public service broadcasters, locked-out CBC workers literally took their talent on the road, through a nationwide tour or caravan to mobilize public support. They also produced the radio equivalent of the strike newspaper by broadcasting "CBC Unlocked" shows on community and university radio stations across the country, including in Toronto, where the CBC has its head office and national broadcast center. This enabled CBC workers, including broadcast personalities who are household names in much of the country, and technicians, to demonstrate their commitment to public broadcasting even as they made use of an important medium to promote their cause. CBC workers, including the "stars," had to sign up as volunteers and pay membership dues to use community stations. As theorists in the autonomist school have repeatedly maintained, creative workers require a lot less than their industrial counterparts to carry out their work when management shuts them out (Terranova 2004). And in this case, in an extraordinary linking up of media workers, their audiences, and alternative media providers, the union succeeded in swaying public opinion to its cause.

The lockout lasted close to eight weeks. Throughout, the workforce remained unified. This is particularly remarkable because workers had been deeply divided over the forced consolidation of bargaining units. They not only patched up their differences but also grew stronger as a single unit. Arnold Amber of the CWA attributes some of this to the effect of being locked out.

> If you want to talk about how to get people together, have them walk around buildings for a couple of weeks and they get to know each other, and they get to see each other as people, not as, "He's a tech and I'm not a tech," or, "She's not a tech and how does she understand what I do?" Well, you get to understand. Everyone faces the same kinds of problems. (Interview with authors 2007)

However, not everyone supported the dominant position taken by the union. For example, some temporary workers welcomed the expansion of contract labor and the kind of flexibility it provides. Others were sympathetic to the budget problems that management faced and either called for more government spending or for a reduced role for the CBC in what appeared to them to be an expanded media environment (Melnicer 2005). But to the surprise of some experts, including those who did not think that the two different types of knowledge workers—content providers and technicians—could get along over the long haul, worker solidarity persisted into the fall. As the lockout dragged on, it threatened the new television season at the CBC, the return of its big Saturday night moneymaker, *Hockey Night in Canada*, the investiture of a new governor-general, and the opening of Parliament. Under increasing pressure, including the threat of losing their own jobs and a September-October decline of about 25 percent of its radio audience, CBC management returned to the bargaining table and settled for far less than the corporation had hoped. The final agreement increased the number of nonpermanent workers slightly, by seventy-five to eighty people, and capped the share of the permanent workforce permitted to be contract workers at 9.5 percent, well below the 14 percent permitted at the BBC. Moreover, after four or more years of contract service, employees can now convert to permanent status (Taylor 2005). In the end, management at the CBC went from an initial proposal to open all jobs at the public broadcaster to temporary contract status, to a position willing to agree that fewer than one in ten would have that status.

Despite what seems to have been a clear-cut victory for the union, the broadcaster continues to look for ways to cut permanent jobs. In April 2006, it announced plans to close its Toronto-based design department and cancelled two youth programs. In 2007, it cut one hundred English-language TV jobs. According to the president of the CMG, Lise Lareau, "TV management is in the final stages of dismantling CBC-made programming. There will be soon nothing left of entertainment television production inside the CBC" (CMG 2006).

In short, the bargaining-unit consolidation at the CBC has turned out to be a striking example of labor convergence at work, a case where a single union is a better voice for workers of all sorts, and where diversity among members contributes to strength and single-minded determination among the leaders. But even with this degree of success, pressures from management to reduce the unionized labor force by outsourcing and job cuts will continue.

Convergence Falls Short in Hollywood

If the consolidated bargaining unit at the CBC is a convergence success story, the relationship between the unions in Hollywood is just the opposite. This section describes an unsuccessful attempt at creating a converged union of cultural workers in Hollywood, the heart of the U.S. entertainment industry and still, despite challenges from Bollywood and other production centers, the film capital of the world. The Screen Actors Guild (SAG) and the American Federation of Television and Radio Artists (AFTRA) have talked about uniting for decades. In 1998, the two finally struck a deal and sent it to members for ratification. AFTRA's members endorsed it; SAG's did not. Five years later they tried again, with a new and different merger agreement. Again, AFTRA's members said yes, this time by a larger margin. So did a majority of SAG members. However, because the SAG vote fell a hair short of the 60 percent threshold required by both unions' constitutions for ratification, the merger was defeated. Says Mathis L. Dunn Jr., the assistant national executive director of AFTRA, "Had we been playing horse shoes or hand grenades, it would have been good enough. But that's not the case" (Interview with authors 2006).

Why would two unions facing an increasingly concentrated set of employers in the film and media industries decide not to merge? To answer this question, we drew on interviews with key participants and access to the primary documentary record. The case is instructive for a variety of reasons. First, there is considerable research on the value of mergers or convergence among trade unions, including in the communication and information industries (Batstone 1984; Katz 1997; Stone 2004). It is important, therefore, to examine why such convergence may not be attractive to workers. Unions are often viewed as bureaucratic institutions, which make merger decisions based largely on a cost-benefit analysis (Chaison 1986). But this case vividly demonstrates that unions are also political institutions with at least formal democratic procedures that may significantly influence decisions about mergers. Specifically, while a merger might help SAG and AFTRA deal with media conglomerates, it might also mean that members would have to give up some of their influence over union decisions, simply because the union has grown larger. Moreover, trade unions are also cultural institutions, and the distinctive cultures associated with a self-consciously craft-oriented group of film workers (SAG) and a union-oriented organization of workers who come out of a broadcasting tradition (AFTRA) also matter. Finally, the story of these two unions provides important lessons for those who would see a natural path to unity among knowledge and cultural workers (Terranova 2004). Unity is not a given simply because workers produce the same product, i.e., digital content. If a merger cannot succeed in a highly concentrated industry between two unions that share a substantial portion of their membership, then perhaps the road to unity among workers in the network society may not be as clear as some would contend or hope.

Given the particularities of the case, it is difficult to draw definitive conclusions that might easily be applied to other merger efforts. However, the SAG-AFTRA story raises questions that knowledge-worker theories, research on the cultural industry workforce, and studies of union mergers generally need to consider. The success that the CWA enjoyed in bringing together content producers and technicians at the CBC, and in creating a bond with their audiences, must be weighed against the consistent failures of these two major Hollywood unions to bridge their own differences and unite.

Chaison (1986) suggests that trade union mergers occur when the incentives to unite outweigh the obstacles. In this case, the incentives were clear: a merger would have put the unions in a stronger position to deal with the challenges of new technology and of an increasingly concentrated media industry. It would have eliminated the need for film, television, and commercial actors to pay dues to two unions. It also would have streamlined the bureaucracy and ended the potential for employers to play one union off against the other. But trade union mergers are not simply a matter of calculating benefits and drawbacks (McKercher 2002). Unions are political institutions, with their own governing and dues structures, bureaucracies, histories, and, especially, cultures. Unlike the corporate mergers that have transformed the communications sector in North America in recent decades, labor union mergers require the approval of the full range of players in each potential partner: the elected leadership, the staff who have day-to-day contact with the membership, and the rank-and-file members. The fact that potential merger partners often have a shared history of competition and conflict, as well as cooperation, complicates matters for trade unions. As a result, mergers between labor unions are often more like the unification of rival states than they are like corporate mergers.

In this case, the particular characteristics of SAG and AFTRA, including their history, culture, sense of craft identity, and the lived experience of members, all helped to derail the unity effort. The failure to merge has had consequences for how Hollywood's cultural workers deal with a range of issues, from new technology to global production.

If ever a compelling argument could be made for a union merger, though, it is the case of the Screen Actors Guild and AFTRA. These unions represent workers in the same industry, doing the same work, often under identical contracts. Between them, they represent the elite of creative workers in the United States. While AFTRA covers a broader range of workers—sound recording artists, radio and television journalists, performers in soap operas, talk shows, reality shows, and game shows—the two unions share an identical core constituency of film, television, and commercial actors. Roughly 44,000 actors belong to both unions, which means 40 percent of SAG's 120,000 members belong to AFTRA, and 60 percent of AFTRA's 70,000 members belong to SAG. Since the early 1980s, under an agreement optimistically called Phase One of Merger—Phase Two was to be the creation of a new governing structure; Phase Three would have brought together the unions' health and pension funds (Prindle 1988, 138)—they have bargained jointly on their major national contracts with the advertising and entertainment industries. Phase One was as far as the unions got.

These unions did not start out as overlapping entities. Originally, they represented distinct groups of workers in distinctly different industries, working on opposite coasts of the United States. What pushed them together was a combination of new technologies and corporate convergence, a combination that continues to blur the lines that separate them today.

The Screen Actors Guild was founded in Los Angeles in 1933. It was not the first union in Hollywood—the Screen Writers Guild appeared a few months earlier—nor was it the first to attempt to organize film actors. Actors' Equity, the New York–based theatrical workers union, ran an unsuccessful recognition strike in 1929. Right from the beginning, SAG was different from a standard trade union. To get around the possibility of individual members being held financially liable for collective action—a quirk of U.S. law that would change with the New Deal—it was set up as a corporation (Prindle 1988, 22). This meant that SAG was a centralized organization, run by a board of directors who would be elected by the full membership and would serve without salary. Voting membership in the union was limited to working actors. One of the unusual features in SAG's constitution was a requirement that strike votes had to be approved by a margin of 75 percent. That bylaw continues to be in effect today (SAG 2007a).

The American Federation of Radio Artists (AFRA) was founded in 1937 in New York, home of the U.S. radio business. AFRA's constitution vested power in the locals, which elected their own boards and chose representatives to the national board. In this respect, it was much more like a traditional North American trade union than SAG. Both unions belonged to the Associated Actors and Artistes of America (AAAA), the American Federation of Labor's umbrella organization for performers' unions. Performance was the common denominator of the two unions, but their jurisdictions were clearly and sharply delineated. Performing on radio was AFRA work; acting before the cameras for later projection on a screen was SAG work. Actors who wanted to perform both on radio and in film had to join both unions, and many did. Indeed, Eddie Cantor, SAG's president from 1933 to 1935, became the first national president of AFRA two years later. Cantor began his career in vaudeville and starred on Broadway, in motion pictures, and on radio. In the earliest days, radio was live work. But in 1941 AFRA asserted jurisdiction over radio transcriptions, or recorded performances that could be broadcast later. This moved AFRA's artists a bit closer to film actors, in that after this point, both radio and film work might be recorded for later use. But clear distinctions remained between the industries, and in the types of work each required.

When the radio networks expanded into the new technology of television, the distinctions between SAG work and AFRA work began to break down. AFRA saw television as an extension of radio. In terms of corporate ownership, the geographic location of the industry, and the way shows were produced and broadcast, that was certainly the case. But television also meant projecting moving images on a screen, and therefore it held the potential for showing filmed programs. Film work was SAG work.

AFRA found allies in other New York–based performers' unions representing theater actors, singers, variety artists, and musicians. Together these unions formed the Television Authority (TVA) and joined with AFRA in an effort to coordinate labor's response to the new technology. As television swept into North American homes after the end of the Second World War, the unions wrangled over jurisdiction. In 1950, the AAAA awarded jurisdiction over television to AFRA, promising that it would become the sole union for radio and television broadcasting artists (Prindle 1988, 76). SAG fought back, appealing to the federal labor regulator for certification elections to allow performers to decide for themselves which union should represent them. SAG won all but one of these elections, and in 1952 AAAA gave up on its attempt to create one big television union. SAG won the right to represent actors in filmed television programs. AFRA was given jurisdiction over live television.

It merged with TVA and took the name AFTRA to reflect its expanded jurisdiction.

This division, which was questionable from the start, began to break down over the next three decades. For one, the television business migrated west, from AFTRA's New York base to SAG's home ground in Los Angeles. Evolving technologies like videotape—which allowed for productions that were neither live nor on film—blurred the lines of jurisdiction. Television's ability to reuse material by showing theatrical films or broadcasting "reruns" of programs launched the unions on long and costly fights for what are called residual payments for members. Meanwhile, actors complained about having to pay dues to two unions for doing essentially the same work. According to Prindle, "Periodically, the notion of eliminating the tangle of jurisdiction by merging all screen actors into one union would gain favor. SAG and AFTRA would consult about it and hire a professional researcher to do a study, there would be a good deal of argument back and forth, and after a while the impulse would peter out" (1988, 77).

Starting in the 1980s, however, the talk became much more serious. Merger discussions were promoted in large part by the introduction of cable television and the videocassette recorder, which threatened both the movie theater business and conventional network television. It also raised the problem of residuals, or payments for reuse of work, to a new level. These new media expanded opportunities for reusing material and decisions would have to be made, within the unions and jointly, on the best bargaining position against an industry that consistently aimed to keep for itself the revenue generated by new media. SAG still had jurisdiction over film and AFTRA over live and taped performances. But these distinctions were becoming less and less meaningful, and there were instances when a producer's decision to change studios—moving a television series from a facility that shot on film to one that shot on videotape, for example—meant changing unions. Because each union had separate contracts, the producers could try to use one against the other. In 1981, the unions took a significant step toward uniting by signing the Phase One of Merger agreement that allowed them to jointly negotiate and ratify all major film and television contracts. The anticipated Phase Two of Merger, which would have brought together governance of the two unions, never appeared. Instead, it became the victim of an internal struggle at SAG.

The Screen Actors Guild not only has an unusual structure for a North American trade union, but also has an unusual political history. This is a union, after all, whose past presidents have included the arch-conservative Charlton Heston and the democratic socialist Ed Asner—not to mention Ronald Reagan, the only trade union president ever to serve as president of the United States. From its earliest days, SAG has had progressive and conservative elements. Indeed, the decision to call itself a "guild" reflected the fact that some of the more conservative founding members were uncomfortable joining a union. At the time of SAG's founding, "guild" was a popular choice among cultural workers in the United States. It harkened back to the artisans' associations of the medieval era, rather than to the factory-based unions of the nineteenth and early twentieth centuries. A guild could be seen as a professional association, and the name suggested exclusivity. In choosing this title for their organization, the actors took their cue from the Screen Writers Guild. A few years later, the directors made the same choice, calling their union the Screen Directors Guild and keeping the "guild" designation when they later changed their name to the Directors Guild of America. The union representing U.S. newspaper journalists made the same choice, naming their organization the American Newspaper Guild. Within a few years, however, the Newspaper Guild had become a full-fledged industrial union, seeking to represent a broad range of workers in the newspaper business. SAG never did. It remained an actors' guild, period.

For its first four decades, SAG's board was dominated by conservatives, though according to Prindle (1988, 8) the struggle between progressive and conservative actors was a "constant but muted" undercurrent in SAG politics. The split between left and right is not the only fault line in the union, however. There have been tensions between the Hollywood, New York, and regional divisions, and between people who want to make their living as actors and people who see acting as a sideline. Some members see a SAG card as a status symbol; others as a meal ticket. In addition, the income structure of SAG members is wildly skewed. Elite members earn millions of dollars per film, while the majority struggle to make ends meet. At any one time more than 80 percent of SAG members are unemployed. Since the 1980s, SAG has developed a de facto party system, an extremely rare feature in North American trade unions. Another communications union, the International Typographical Union, was the only other North American trade union with an institutionalized two-party system (see Lipset, Trow, and Coleman 1956).

SAG's party system is neither formal nor institutionalized, but it is nonetheless influential. The names (and at times the composition) of the parties change, depending on the issue grabbing their attention, or on the views of the caucus of members promoting or contesting a particular issue. Competing slates of candidates for national office have become a regular feature in SAG elections and this has had an enormous impact on attempts to merge with AFTRA. In 1980 a group known as the Caucus of Artists for Merger, upset with the leaders of both unions and angry over what they saw as a botched strike that year, set out to gain power and influence in both the SAG and AFTRA governing structures, hoping to use that platform to bring about a merger. They won control of AFTRA's Los Angeles local the next spring, convinced the well-known television actor Ed Asner to run for president of SAG a few months later, and in that year's election took control of the SAG board. Their hopes for a speedy merger were derailed by the creation of a conservative coalition, Actors Working for an Actors' Guild, whose members objected to Asner's radicalism in particular, and to the direction in which SAG was heading in general. The two factions tore at each other for much of the decade, ultimately sidelining the AFTRA merger talks in 1988. The SAG-AFTRA merger effort was knocked right off the agenda by a vicious fight over a tangentially related merger between SAG and the Screen Extras Guild, which was to be a prelude to the larger merger with AFTRA. That deal was rejected by the SAG membership twice, the second time more decisively than the first (Prindle 1988).

In the early 1990s, the unions decided to try again, going at things more slowly this time but making steadier progress. Finally, in 1998, they made a deal. The agreement called for a full merger of the two unions into a new body, to be called SAG/AFTRA, whose structure would reflect the traditions of both unions. It would have a strong national board of directors (a SAG feature) and thirty semiautonomous locals (an AFTRA feature). The new union would have a new dues structure, but the two unions' health and pension plans would remain separate (Robb 1999a). In promoting the merger to their members, the leaders of SAG and AFTRA argued that the merger would end jurisdictional disputes between the two unions, including a looming battle with the media conglomerates over revenues from high-definition television. It would also streamline operations, and give the union more power at the bargaining table.

The merger agreement provoked a strong and vigorous reaction under the banner of a group called Save SAG. This largely conservative opposition group, a faction of a broader coalition known as the Performers Alliance, dismissed the prospect of a jurisdictional dispute as a scare tactic. The group also objected to the idea that dues would increase after the merger, and they questioned the decision on health and pension plans. But perhaps the biggest success of the Performers Alliance was in convincing SAG members to embrace a narrow, craft-based view of trade unionism—to choose the "purity" of an actors' union rather than merging into a union with broadcasters and musicians (Cooper 2001).

Ballots went out to members in November 1998, with polls closing in January 1999. Both unions' constitutions required a merger vote to pass by a margin of 60 percent. When the votes were tallied, 67.6 percent of AFTRA voters supported the merger; but only 46.5 percent of SAG members voted for it (Kiefer 2003; Robb 1999b).

Soon after, SAG went through another period of political turmoil. Its pro-merger president, Richard Masur, was replaced by the conservative William

Daniels, who promised to get tough with the employers and presided over a divisive and combative two-year term that included the longest strike in the union's history, the six-month SAG and AFTRA commercials strike in 2000. Assessments a year later suggested that SAG's hard-line approach prolonged the dispute unnecessarily (Cooper 2001). Daniels decided not to seek reelection in 2001, throwing his support to Valerie Harper. She lost the election to Melissa Gilbert, but the result was nullified because of balloting irregularities. Gilbert won the rerun election. A moderate, Gilbert headed a coalition known as Restore Respect, whose platform included support for a merger with AFTRA. Negotiations were on again, and in 2003 they resulted in a new agreement.

This deal, based on a set of principles drawn up by a joint committee of the two unions, tried to avoid some of the stumbling blocks that tripped up the previous merger. To recognize that SAG members didn't want to lose their identity or their traditions, the deal proposed an affiliation rather than a full merger. The new organization would be called the Alliance of International Media Artists. It would have three autonomous, self-governing affiliates: SAG, which would cover actors of all sorts; AFTRA, which would cover broadcasters; and the American Federation of Recording Artists, which would cover singers and recording artists. Each affiliate would have its own board and officers. Again, the health and pension plans would remain separate, but the trustees of both would look into the possibility of consolidating them (Connell 2003; Kiefer 2003).

In announcing the details of its consolidation plan to members, *AFTRA Magazine* (2003, 5) pointed out that one of the "common perceptions" in the 1999 merger result was that there was "no imminent or compelling need to take action." This time, it argued, things were different: "The emergence of digital production has rapidly escalated the jurisdictional conflict. This serves to: 1. Drive down terms and conditions for performers. 2. Divert resources from critical union initiatives."

If new technology was a compelling factor in favor of a merger, so was industry consolidation. The same issue of the magazine summarized the changes: the number of major employers of performers and broadcasters had declined from twenty-six in 1985 to six, five of them in the television and motion picture industry; eight studios controlled 88 percent of all domestic theatrical production and distribution; five record companies controlled 84 percent of the recording industry; cable revenues increased from $5 billion in 1990 to $35 billion in 2002, and the cable market was controlled by five companies; two cable system owners, AT&T and Time Warner, accounted for 40 percent of all cable households; two companies owned more than fourteen hundred radio stations and one, Clear Channel, was comprised of what used to be seventy separate companies. The magazine put it this way: "While AFTRA and SAG have been *thinking* about consolidating, our employers have actually done it. The industry *already* has consolidated" (2003, 4; emphasis in original).

As with the previous merger vote, the campaign was heated. A reinvigorated Save SAG argued that the plan would result in a loss of direct fiscal control over SAG's budget, and that the actors union would be reduced to a "glorified committee" of a larger union (Kiefer 2003). In an interview with the *New York Times*, former SAG president Daniels brusquely dismissed the idea that a merger would make SAG stronger: "AFTRA is a crummy little union, and they're undercutting our contracts, so we should join them? To us it's a very, very bad deal" (quoted by Connell 2003). The opponents also made much of the uncertainty about the future of the pension and health plans (Connell 2003)—a tactic that frustrated members of the committee that negotiated the deal. Matt Kimbrough, AFTRA's national recording secretary, explained that because the plans were created and governed separately from the unions, "you couldn't even begin the process of merging the plans until you actually merged the unions" (Interview with authors 2006). But this meant that merger advocates could not offer clear answers to questions about what the merger would mean for actors' pensions, a situation that opponents to the merger used to fan actors' fears about the future.

The voting took place in the summer of 2003. AFTRA members overwhelmingly endorsed the merger, with 75.8 percent of votes in favor. At SAG, the deal lost by the narrowest of margins. The merger needed 60 percent of votes to pass; it got 57.8 percent (Kiefer 2003).

Pamm Fair, the deputy national executive director of SAG, says post-election polling of SAG members found that the vast majority wanted a merger; "they just didn't like *that* plan." They also objected to the idea that the new union would have a different name. "Our people said, 'What, are you crazy? We have worldwide branding with the name we have now. Why would you think about changing it?'" The pension and health-care issue also resonated with members, especially because SAG's plans are comparatively healthier—and less expensive for members—than AFTRA's (Interview with authors 2006).

Kimbrough, who is also a member of SAG, said the opponents of the merger came up with a short and simple slogan to sink the deal: "If you don't know, vote no." It worked brilliantly. "The members were sold on the notion that if you vote no now, there's going to be another vote later, when there was never any intention from the people trying to defeat it that there be a merger agreement." He agreed that the name of the new organization was an issue, as it had been in 1998. Hollywood is, after all, SAG's town, and the SAG name is known throughout the world. Kimbrough indicated that if the two sides had simply agreed to call the new union SAG, the merger might have passed. But that would have been unacceptable to AFTRA, since it would imply an absorption by another union. Leslie Simon, director of legislative and public affairs for AFTRA's Los Angeles local, added another reason for the failed merger: "Fundamentally, there was a lack of recognition of how the industry has changed, of why it is that two unions are no longer acceptable and no longer can do the job for actors" (Interview with authors 2006). Finally, as in 1999 the promerger side was unable to overcome the opponents' appeal to craft solidarity—the notion that as performance artists, actors shouldn't be comingled in a union with broadcast journalists or musicians. Ron Morgan, president of AFTRA's Los Angeles local and a SAG member as well, said this argument was strictly an emotional one. But when all was said and done it "resonated stronger than anything else" (Interview with authors 2006).

After the merger was defeated, each union had to chart a new course based on maintaining the status quo of two unions with a history of working together, rather than determining how to work as one. SAG elected a new president, Alan Rosenberg, whose Membership First coalition currently holds a majority on the national board. Rosenberg opposed the AFTRA merger; instead, he would prefer to see all actors come to SAG. A political liberal who was an antiwar activist during the Vietnam War, worked in the civil rights movement, and is an advocate for social justice, Rosenberg pledged—as the conservative Daniels did—to get tough with the employers. He told the *Los Angeles Times*: "Until now, the studios have looked at how fractured we are as a union and haven't taken us very seriously. So my philosophy is that we have to take an adversarial position and stand strong and never leave the negotiating table until we make progress and prevail" (Goldstein 2005). Rosenberg, who won reelection in 2007, presided over the firing of the union's executive director, a key player in negotiating the failed AFTRA merger. Under his leadership the union has also taken a toughened stance on so-called financial core or fi-core members. The term refers to members who resign their SAG membership and withhold the portion of dues the union spends on political activities, but are still permitted to work on union jobs. Financial core membership in U.S. unions became legal following a 1988 Supreme Court ruling on a case involving the Communications Workers of America. Typically, SAG workers who opted for "fi-core" status could return to full status with SAG as desired. But the union has promised to make it harder to move in and out of the union, thereby sending a message to its members that it plans to enforce union discipline, particularly around initiatives like Global Rule One that attempt to assert the union's authority nationally and internationally (McNary 2007a).

Rosenberg has also begun to work more closely with two other Hollywood unions that have fought their own internal battles: the Writers Guild of America (WGA) and the Directors Guild of America (DGA). "If we could present a united front with the WGA and the DGA, we'd be a lot stronger," he told the *Los Angeles Times* (Goldstein 2005). Since late 2005, the writers' and actors' guilds have been campaigning together for tougher rules on product placement, which the unions see as "stealth advertising."

In March 2007, SAG began meeting with the Writers Guild of America to prepare for negotiation for a new television and film contract with the Alliance of Motion Picture and Television Producers. AFTRA, normally SAG's partner in these talks, was excluded from the meetings. In the fall, negotiations between the WGA and the producers broke down, largely over the ever-recurring issue of residual payments. What, if any, compensation should writers receive for work that is increasingly repackaged for the digital world, especially the Internet? With the producers proposing to end the system of residuals and develop new forms of compensation, and the writers determined to press for expanded residual payments, it came as no surprise that the WGA took to the picket line on November 1, 2007, in a labor action that would have significant implications for the television and motion picture workforce. Both SAG and AFTRA supported the writers, often joining them on the picket line and providing individual contributions to the WGA strike fund. The strike, which lasted three months and cost Hollywood an estimated $2.5 billion (Guider 2008), was resolved with a deal that promised writers a stronger foothold in the online world. In the first two years of the three-year contract, writers would get a maximum flat fee of about $1,200 for programs streamed online. In the third year, however, they would receive 2 percent of a distributor's gross, a key union demand. Other provisions include increased residual payments for downloaded movies and TV programs (Elber 2008).

Following the failed merger with SAG, AFTRA undertook a major restructuring. It decided to move its national headquarters from New York to Los Angeles, home of its largest local, though some key departments, including news, will remain in New York. It embarked on a series of structural changes—reducing the size of its executive board, for example—that are aimed at cutting costs, reducing its deficit, and increasing its revenue (*AFTRA Magazine* 2004–2005, 2006; Horwitch 2005). It elected a new president, Roberta Reardon, after John Connolly stepped down in March 2007 to become national executive director of Actors' Equity Association. AFTRA also began discussions with the International Alliance of Theatrical Stage Employees (IATSE), the union that represents mainly technical workers in film, video, and theater, to develop a closer relationship. As the president of IATSE put it, the goal is to "work to ensure that all our members across all of our disciplines speak with one strong voice in the entertainment workplace" (Salomon 2007). In addition, delegates at AFTRA's 2008 national convention voted overwhelmingly in favor of seeking direct affiliation with the AFL-CIO, rather than continuing to be affiliated through Associated Actors and Artistes of America, the umbrella organization for performers' unions that also includes SAG. AFTRA president Reardon described the significance of the move this way: "Directly affiliating with the AFL-CIO would place AFTRA as an independent and autonomous union within the Federation, with all of the rights and responsibilities of other direct affiliates and with complete control of our own destiny in determining our relationships, partnerships, and alliances with other unions affiliated with the AFL-CIO" (Reardon 2008).

SAG and AFTRA have continued to work together in several areas, though their relationship has at times been strained. The unions successfully negotiated a contract in 2005 covering voice-over actors featured in video games, an area that was not included in the 1981 Phase One agreement on joint bargaining. Both are also active in the formation of the AFL-CIO's new industry coordinating committee, which covers unions in the arts, entertainment, media, and telecommunication industries (AFL-CIO 2005; Eggerton 2005). The federation says the committee's goal is to build more power for workers by undertaking collaborative initiatives in four areas: organizing, collective bargaining, contract standards, and public policy. Pamm Fair of SAG, who works on policy and strategic planning, said she hopes the committee will serve as an arbitrator for jurisdictional problems that may come along. Kimbrough said AFTRA is hoping to play a pivotal role in the creation of the council and is keen to see it succeed. But he was not certain about SAG's commitment to the idea. "Right now I think that in a lot of these questions, the guild is a wild card," he said (Interviews with authors 2006).

Perhaps the most significant fallout from the failure of convergence is that the unions now find themselves competing with each other for contracts in some areas, a situation that leads to anger and frustration on both sides. A host of new devices and media have arrived since the last merger vote—podcasts, satellite radio, iPods with video screens, interactive blogs, cellular telephones with video, and so on. These devices have all but obliterated the lines between various types of performance, and pose the most serious challenge yet to jurisdictional divisions. In January 2006, AFTRA president John Connolly set out his union's stance on digital technology, recording a podcast to do so: "This message . . . [is] an acknowledgement of AFTRA's readiness and resolve to organize the digital markets of the 21st century—proof positive that AFTRA is the future in media" (AFTRA 2006). Shortly after that, it announced that two new prime-time series shot in digital format were under AFTRA contract (Miller 2006). Because SAG also claims that its contracts cover digital production, this means that the two unions are now competing for digital work. Both are aware that competition tends to benefit the employers rather than the workers. As AFTRA's Kimbrough put it, "The only way a union competes with another union for a job is to offer the producers a better deal. That's a hard truth and it's not good for the members" (Interview with authors 2006). But each union feels it has no choice.

In late 2007, as the unions were preparing for joint negotiations on a new film-TV contract, relations between the two continued to degenerate. In a twelve-page letter sent to members in October, SAG's executive director Doug Allen accused AFTRA of poaching contracts on cable shows and undermining the joint bargaining position at the forthcoming negotiations with the Alliance of Motion Picture and Television Producers. AFTRA's Reardon fired back, accusing Allen and SAG leaders of manipulative attacks and fear-mongering. She also asserted that the criticism from SAG had an ulterior motive: "There is another thinly veiled campaign supported by some SAG paid staff and others to raid another union—the cardinal sin of treachery within the labor movement" (McNary 2007b). Days later, AFTRA accused SAG of breaching the long-standing Phase One joint bargaining agreement by changing the way SAG members of the bargaining committee would vote. She called for SAG to reject the new plan for bloc voting on the committee and revert to the way the committee operated in the past. AFTRA also threatened to go it alone in the negotiations (McNary 2007c). As the writers' strike was ending in early 2008, SAG attempted to make peace with AFTRA. At an emergency board meeting, it scratched the bloc voting idea (Cieply and Barnes 2008). Nonetheless, in March 2008, AFTRA decided to suspend its joint bargaining agreement with SAG.

Another area where the failure to merge is likely to have consequences is in the campaign against what Hollywood likes to call runaway production—the production of films and television shows in lower-cost, incentive-rich, non-Hollywood locales like Canada. As we describe in chapter 7, SAG's approach has been to build a series of coalitions to address the issue in several different ways. With the Directors Guild, it commissioned a groundbreaking economic analysis of the problem. It then helped create a nineteen-member coalition known as the Runaway Production Alliance to fight for federal and state legislation to subsidize domestic film production. The alliance includes not just trade unions, but professional and trade organizations, as well as groups representing marketers, distributors, promoters, talent agents, and musicians—in other words, anyone in Hollywood with a stake in the issue. SAG has also extended its jurisdiction beyond the U.S. border through its Global Rule One campaign, which maintains that SAG members can only work for producers who sign SAG agreements, regardless of where they film. AFTRA participates in the Runaway Production Alliance and has issued public statements in support of the extension of Global Rule One. But since the failure of the merger vote, it has been considering whether to continue the lobbying effort against runaway production at the same level as before. The union is not sure that it is worth the time, money, and effort. "We've spent a considerable amount of energy and capital in support of these bills," Kimbrough explained, "and that has gravitated to the benefit largely of the Screen Actors Guild. There has not been a terrific amount of growth in AFTRA work as a result of these bills" (Interview with authors 2006).

SAG also worked with a different coalition, the Film and Television Action Committee (FTAC), to prepare a complaint to the Office of the United States Trade Representative against Canadian film subsidies. AFTRA has had no interest in this idea. Kimbrough said, "Politically we have voted it down every time it comes up in our boardroom. We regard it as counterproductive" (Interview with authors 2006). In September 2007, FTAC filed the complaint with the trade representative's office, seeking to have Canada's subsidies declared a violation of trade agreements. Six weeks later the trade representative's office rejected the petition, saying it found that neither the economic data nor the legal arguments set out in the petition supported the group's contention that Canadian film subsidies are unfair.

Ron Morgan, the president of AFTRA's Los Angeles local, said that focusing too much attention on runaway production misses the larger issue confronting Hollywood unions. "The real runaway production issue, to me, is not that the work is going to Canada, but that it's going nonunion. That's runaway production as far as I'm concerned" (Interview with authors 2006). If Morgan is right, the solution to the problem of the increase in nonunion work is to capture—or in some cases, recapture—the work. And this is where the lack of labor convergence may play a critical role.

We began this chapter by suggesting that the twin challenges of new technology and corporate concentration are prompting worker organizations to explore new ways of increasing labor power. Few places offer as vivid an example of both phenomena as Hollywood. Between them, corporate consolidation and technological convergence have undermined existing union jurisdiction, opened up new areas of nonunionized work, and tipped the balance between labor and management sharply in favor of the employers. A merger between SAG and AFTRA would have helped the members of both unions. It would have ended the competition for work between SAG and AFTRA, a competition that chiefly benefits the employer. Individual actors would pay only one set of dues, not two. The organization resulting from the merger would be simultaneously more united and, with AFTRA's nonacting membership, more diverse. In short, the new union would have been more like a convergent union than a craft union.

In the fall of 2007, a group of 240 supporters of a SAG-AFTRA merger launched a public relations campaign to revive the push to combine the performers' unions. The group included former SAG president Melissa Gilbert, a long-standing advocate of a merger, three other former SAG presidents, and the current president of AFTRA (McNary 2007d). In the short term, talk of reviving a SAG-AFTRA merger would appear to be largely a case of wishful thinking. In the longer term, shoring up their mutual relationship while building strong bridges to other cultural worker organizations, in Hollywood and beyond, may be critical to whether SAG, AFTRA, or some combination of the two will thrive or merely survive. The necessity for concerted action was underscored in July 2007, when a united front of the major Hollywood producers, the powerful result of industry convergence, jointly called for the end of the entertainment industry's decades-old system of paying so-called residuals for the reuse of movie and television programs after their initial showings (Barnes 2007). Their alternative would be to set up a fund to be shared with performers, writers, and directors, but only after the companies recover their costs. Emphasizing that convergence is driving this proposal, the chief executive of Warner Brothers, Barry Meyer, explained: "There are no ancillary markets anymore; it's all one market." Drawing on decades of experience, the unions are questioning whether they can trust the companies to provide an accurate accounting of costs and profits. But whether they can back that conclusion with an effective counterattack will greatly depend on how well they will be able to work together (Cieply 2007).

Chasing a Convergent Company: TWU versus Telus

The previous cases have shown that labor convergence is a complex phenomenon. Sometimes it works; sometimes it doesn't. The next case shows that it is also not an all-or-nothing phenomenon. Partial convergence is possible, and it can have both positive and negative consequences.

Over the last two decades, Telus has grown from an Alberta-based telephone company to become Canada's second-largest telecommunications company, with $8.6 billion in revenue in 2006 (Telus Corporation 2006). It employs about thirteen thousand people, most of whom are members of the Telecommunications Workers Union, based in British Columbia. Much of Telus's growth came through convergence, as the company transformed itself from a firm that emphasized traditional wired telephony to one that offers residential and business telecommunications and Internet service, as well as TV service in selected markets in Alberta and British Columbia. The TWU's growth mirrored the company's, expanding geographically and in the types of members it represented, as the company expanded. These developments contributed substantially to a labor dispute that escalated over four years and resulted in a 2005 strike/lockout that raised significant questions about the meaning and efficacy of labor convergence.

The provinces of Alberta, Saskatchewan, and Manitoba had a long tradition of provincial ownership of infrastructure, including telecommunications. This began to end in the 1980s, with the arrival of neoliberal governments that promoted private interests and competition in the marketplace, and backed away from public service and public interest regulation. Telus was created to take advantage of newly introduced telephone competition in Canada. It was formed in 1990 following the reorganization of the Alberta Government Telephones Commission, the dominant telephone company in Alberta. Telus became AGT's parent and completed its acquisition of shares from the Alberta government a year later. In 1995, Telus bought ED TEL—Edmonton, Alberta's municipal telephone company—and three years later it merged with BC Tel, a privately owned company that was the primary telecommunications provider in British Columbia. These mergers and acquisitions made Telus the dominant telecommunications provider in western Canada. The company didn't simply grow larger; it also expanded the types of services it sold, moving well beyond traditional wired telephony into wireless and Internet services. In 2000, Telus moved east, into the central Canadian heartland of Ontario and Quebec. It obtained a controlling interest in QuébecTel, a local service provider in southern Quebec, and acquired Clearnet Communications, a digital mobile telephone provider. This allowed the company to provide wireless telecommunications services with national scope.

The Telus merger with BC Tel meant that it had to face one of the strongest and most militant unions in the province, if not the country. Like other telephone workers unions in North America, the TWU began as a company union. It gained independence in the 1940s as a result of legislation discouraging company-organized unions, similar to the U.S. Wagner Act of the 1930s. Then known as the Federation of Telephone Workers of British Columbia, the union had to decide early in the postwar years whether to expand its interest beyond the province and work toward the creation of a national union, or deepen its power base in British Columbia. Some union leaders advocated (and campaigned actively for) affiliation with the Communications Workers of America, which was trying to expand into Canada. Others dismissed it as a "foreign union," a characterization that was echoed decades later in the fight over CBC bargaining-unit consolidation. Three separate votes in the late 1940s and early 1950s, each dealing with a slightly different variation on the idea, rejected affiliation (Bernard 1982). What it couldn't achieve by negotiation, the CWA attempted to achieve by force. By then an international union with Canadian members, the CWA launched a raid on the

B.C. union in the early 1960s. The union fought it off and the CWA was forced to concede defeat.

The 1960s were a period of growing militancy for many trade unionists, especially in British Columbia. The B.C. telephone workers began to get tough with their employer, launching their first strike in fifty years in 1969. The union renamed itself the Telecommunications Workers Union in 1977, in recognition that the traditional telephone business was changing, and continued to engage in a series of battles with BC Tel over job security, work assignment, and control in the workplace. During the most hard fought of these, in 1981, the workers took over the provincial telephone exchanges to demonstrate their ability to run the company without management. (For a full account see Bernard 1982.) The battles with BC Tel at times assumed epic proportions. Bernard notes that the 1981 dispute was "a confrontation lasting 536 days, including a four-month selective strike, a seven-day occupation in Nanaimo, a five-day province-wide occupation of telephone exchanges, a one-day general strike in Nanaimo and the intervention of the federal labor minister, the provincial labor minister and half a dozen mediators" (1982, 222). The union's ferocity was rewarded. The union's current research director notes that the TWU had arguably the best protection against contracting out in North America; significant control over the workplace and the working lives of members; and the ability to play "a uniquely proactive role with BC Tel management" (Shniad 2007, 301). BC Tel may not have liked the situation, but the provincial government and the phone company chose to make peace with the union rather than disrupt a vital public service.

The merger of BC Tel and Telus challenged all that. Telus embodied the corporate world's interest in pushing the limits of a market-based telecommunications industry. Making peace with a union committed to public service telecommunications and a significant amount of worker control over the workplace would not work. The result was a four-year war between the company and the union.

The TWU was not the only union in the newly merged company: the International Brotherhood of Electrical Workers (IBEW), the Communications, Energy and Paperworkers Union of Canada (CEP), and the Civic Service Union also represented Alberta's Telus workers. Realizing that the TWU posed a formidable challenge, Telus initially proposed to restructure the newly merged company into four divisions: wireline, wireless, Internet, and business information technology services. Corresponding to these divisions would be four units of workers. The TWU would be limited to the unit identified with wireline service, the service expected to decline as a share of the company's business. This would keep the TWU from representing workers in the fast-growing wireless and Internet services divisions. The TWU saw this as an attempt by the company to minimize the potential damage that a militant union might inflict. The company would then be free to locate more co-operative union members in other units, and perhaps free them from union representation entirely.

But in order to make the case for this proposal, Telus had to reverse its traditional position on convergence. In the past, Telus had described convergence as an inevitable process that would eventually connect all technologies in a seamless network—or the telecommunications equivalent of Bill Gates's friction-free capitalism (Gates 1999). Now, however, Telus argued that there was such a significant distinction in technology, services, regulation, and customers between what in 1998 it called voice, radio (meaning mainly wireless), Internet, and business IT services that each required a distinct business division and a distinct workforce. Moreover, it made an additional geographic distinction, claiming that TWU representation was inappropriate outside the province of British Columbia because that was the province where its members historically did their jobs (CIRB 1999). In essence, the position Telus set out in its filing to the CIRB demonstrates the discursive significance of convergence. Convergence does not simply provide an empirical description of a process, but serves as a political tool for carrying out a company policy. According to this view, convergence exists when the company aims to create a unified brand that identifies it with technological progress. But it does not exist when the acknowledgment of convergence means permitting a well-organized and historically militant trade union to represent workers throughout the company.

The TWU opposed this position, maintaining that the divisional and bargaining-unit structure which Telus proposed was largely set up to restrict its members and curtail the union. In addition—and drawing on earlier claims made by the company itself—it insisted that differences in services and regulatory regimes were rapidly eroding (Mosco 1999). The CIRB, whose earlier rulings showed it was broadly committed to the convergence principle and to bargaining-unit consolidation in converged companies, ultimately agreed with the union. It determined that TWU members should have access to jobs throughout the company, and that the TWU had the right to be the sole bargaining agent at the company. In principle, this marked a significant defeat for the company. However, the CIRB decided to take a conservative approach to implementation by permitting existing contracts with other unions to remain in effect until such time as the TWU and Telus reached an agreement on a contract covering all workers in British Columbia and Alberta, a decision that slowed the TWU's momentum toward consolidating its new and expanded membership. Meanwhile, the company adopted a new strategy that involved a similarly political use of convergence.

The next stage in the struggle, beginning in 2001, saw Telus create a separate company, Telus Mobility, to house its wireless operations, including the newly acquired Clearnet wireless. Again, Telus sought to keep the TWU out of the new company and the union petitioned the CIRB to fight this action. This case had a precedent that ultimately would not bode well for Telus. In an earlier case, Island Telecom, the major telephone company in Canada's smallest province, Prince Edward Island, sought to create a wholly owned subsidiary, Island Telecom Advanced Solutions (ITAS), to operate its Internet services. By sequestering its fast-growing Internet services in a new subsidiary, Island Telecom thereby hoped to keep the CEP union from representing workers at the new company. The union took the case to the CIRB and argued that given convergence, such a division was artificial. The company claimed, on the other hand, that convergence did not hold in the relationship between telephone and Internet services. The CIRB ruled that convergence did hold in this case and that Island Telecom and Island Telecom Advanced Solutions were one employer for the purpose of bargaining. It concluded that the CEP had the right to represent workers at ITAS (CIRB 2000c). This ruling was significant. Had the CIRB ruled against the CEP, it would have established a precedent that telecommunications firms could set up separate subsidiaries in different areas of communication and information technology in order to shut out trade unions.

In spite of the Island Telecom ruling, however, Telus sought to create a separate subsidiary in Telus Mobility and justified the decision before the CIRB with the claim that differences between wireline and wireless technology, and between the regulatory regimes set up to govern them, were significant enough to justify a division that would bar the TWU from representing wireless workers at the company. But once again, convergence bit back. The CIRB decided against the company, giving the union the right to represent workers throughout the company (CIRB 2004a).

What it failed to achieve before the CIRB, Telus next sought to accomplish at the bargaining table. The company and the union had vastly different visions of telecommunications. Telus had built the company by expanding and integrating a variety of telecommunications technologies and services from British Columbia to Alberta, and eventually into Ontario and Quebec. Its goal was to compete nationally in a regulatory environment governed by market principles: that is, the market should govern the price of telecommunications services and the price and supply of labor. The union sought to retain the strong position it had attained through years of struggle with BC Tel, and to extend it to Telus's new markets and services. The TWU supported a publicly regulated telecommunications system, and some in the union called for a return to public ownership. The TWU believed in—and had previously experienced considerable success in fighting for—a social contract that provided full-time, secure, and well-paid jobs for its members. Both the company and the union used convergence to make their positions before the CIRB and both held firm on contract demands. Although it continued to lose before the CIRB, Telus appeared to be in no hurry to incorporate the board's decisions in a new contract. More likely, it hoped to have success across the bargaining table because Telus knew the TWU was facing significant challenges from the inside.

The union had succeeded in its plan to become the sole bargaining agent for unionized workers at Telus. But winning the right to represent the workers in Alberta, and the Clearnet workers in Ontario and Quebec, was one thing; achieving the kind of labor convergence that succeeded in the CBC case was another matter. Geography presented a special problem: the leadership and the long-standing membership of the TWU were in British Columbia, a province with high union density, a long history of trade unionism, and, in the arena of provincial politics, a strong social democratic party. The new members were in different provinces, including neighboring Alberta—the least labor-friendly province in Canada—but also Ontario and Quebec, thousands of miles away from the TWU's west coast base. The Clearnet workers had no union representation and included numerous French-speaking workers from Quebec. The TWU had no experience with the politics and culture of Quebec labor struggles.

One of the things that made labor convergence possible at the CBC was the close working relationship between the bargaining units. The CEP technicians and guild editorial workers knew each other well from decades of working together, side by side. Individual members of these two unions at the CBC had personal relationships and trusted each other professionally, even when their unions were fighting over bargaining-unit consolidation. At Telus, this was not the case. The Alberta Telus workers were trade unionists with their own traditions, collective bargaining agreements, and affiliations. Some Alberta union members preferred their previous union affiliation—with the International Brotherhood of Electrical Workers, for example—and were suspicious of the TWU. And although the workers now belonged to one union, they were still working under different conditions, negotiated under the previous labor regime. Finally, the unions at the CBC had some time to sort out a new relationship before collective bargaining began. The TWU did not have that luxury; it had to work on collective bargaining at the same time as it was trying to consolidate its membership and reinvent itself as a national union.

All the TWU's agreements expired on December 31, 2000, and the union continued to work without a contract. Over the next several years, the company and the union fought things out in a lengthy series of court challenges and CIRB hearings. Eventually, in 2004, the CIRB issued a ruling determining that Telus had committed several violations of the labor code and should offer binding arbitration to settle the dispute (CIRB 2004b). When the union signaled its support for the idea, the company took the case back to the CIRB and got the board to back away from its support for binding arbitration. In April 2005, the company escalated the dispute with what amounted to a "soft lockout," suspending the remission of union dues and completely ignoring the established grievance processes. Telus announced that unless a contract was reached in three months, it would unilaterally impose one (Shniad 2007). The TWU responded with its own actions, including refusing overtime, working to rule, and slowing down the repair and installation process by refusing to permit its members to work out of a home base. It also began time-consuming study sessions and, in some locals, actually occupied company facilities. The company reached the brink when it announced that it would impose its own version of the contract on July 22, 2005. The union responded by pulling its workers off the job on July 21.

The lockout/strike brought together many elements of the classic information age labor dispute. The union launched an extensive media campaign, which called on residential and business customers to give up their Telus services. It also convinced Vancouver's Pacific Newspaper Group, a Canwest division that owned the city's two major daily newspapers, to refuse to accept advertisements from Telus. The company brought in strikebreakers from other parts of the country and outsourced some of its work to call centers in the Philippines. It also blocked access to a website, Voices for Change, set up by a Telus employee, which the company claimed contained photos that might intimidate or threaten people involved in the dispute. In the course of blocking Voices for Change, Telus also denied access to 766 sites unrelated to the labor dispute that happened to use the same server. This action, extraordinary for an Internet service provider, attracted widespread attention. According to the OpenNet Initiative—a partnership between units at the University of Toronto, Harvard Law School, and Cambridge University—Telus kept its subscribers from accessing an engineering company, an Australian site promoting alternative medicine, a Colorado firm that recycles electronics parts, and an organization that raises funds for breast cancer research (Barrett 2005).

Meanwhile, the company was able to exploit the climate of hostility to organized labor in Alberta, as well as the incomplete degree of labor convergence between the Alberta and B.C. members of the TWU. By September, more than half of the TWU's Alberta members had returned to work, thereby allowing the company to restore service to close to normal levels. The TWU enjoyed the support of other unions, including the CEP workers at the Vancouver newspapers who pushed for the boycott of Telus ads, but it suffered from one shortcoming in its approach to convergence: it was limited to the telecommunications arena. Certainly, the union had expanded into the wireless and Internet service occupations, a version of labor convergence, and had expanded geographically. But unlike the CEP or the CWA, it did not have units in media or other industries from which it might draw in this difficult time. The TWU's convergence was at best partial and the union suffered as a result.

With so many of its members returning to work, the union stepped up its return to the bargaining table. It is always difficult for outsiders to determine precisely what constitutes victory in a labor dispute, particularly in the knowledge industries where control and security of employment typically loom as large as, or even larger than, wage issues. In this case, the history of militancy at the TWU makes an assessment more difficult because the union was less inclined than other unions to believe that it needed to give up on any demand. For the TWU, victory meant maintaining the strong position it had enjoyed with its former employer, BC Tel. The two sides reached a tentative agreement in October, but the membership voted it down by a razor-thin margin—4,540 against ratification and 4,487 in favor. The union's leadership was forced back to the bargaining table to reach a second deal. When this one was put to the members, 67.3 percent voted yes (Shniad 2007). The contract gave workers a 2 percent annual raise for each of the five years of the contract, $10 million in a pay equity settlement, limitations on contracting out, guarantees that TWU members would work on the company's Internet television project, maintenance of the union pension plan, and an affirmation that the TWU was the sole bargaining unit for the Telus workforce. Nevertheless, there was little satisfaction at the union. According to its research director:

> No one is enthusiastic about the new contract. A great deal of emotional energy is focused on the things that have been lost. This defeat has generated understandable anger and frustration among TWU members. Such feelings are to be expected. Unfortunately, there is often a tendency in such circumstances to engage in personal recriminations in an attempt to lay blame for a bitterly disappointing outcome. (Shniad 2005)

One area that continues to raise concerns centers on the meaning of limited contracting out. The contract promised that no full-time jobs *directly* related to contracting out would be lost. Union members fear, with some justification, a steady erosion in jobs as work is outsourced to contract workers or overseas, and TWU jobs loosely defined as *indirectly* related are lost. This would certainly follow an increasingly common pattern of outsourcing work in the legal, medical, media, finance, and telecommunication sectors (Mosco 2005). Under the old agreement with BC Tel, any normal or regular work could not be cut without the permission of a standing committee comprised of management and labor. Under the new agreement, Telus committed to job protection but provided no guarantees of union involvement in such decisions. Making matters more difficult for the union, the assessment of the contract varied depending on whether workers came from British Columbia or Alberta. So while the union may have used convergence to bite back at Telus on several occasions, it is likely that—short of following the ultimate convergence strategy and merging with the CEP, the CWA, or another Canadian union—the last bite will be taken by Telus.

In an ironic postscript to the long struggle over convergence, Telus announced shortly after the contract was settled that it would merge Telus Mobility with the parent company in order to take advantage of convergence between technologies (Wong 2005). This decision completely reversed the position it set out earlier before the CIRB, where it argued that there was no technological convergence between wired and wireless telephony and that its decision to keep them separate had nothing to do with its desire to keep the TWU out of wireless. This change of direction would appear to validate the CIRB's rejection of that view. It also offers evidence that convergence can be a political and rhetorical phenomenon as well as a technological one.

The Complexities of Convergence

It is certainly important to examine the convergence of technologies, industries, and corporations. It is also valuable to examine convergence as a cultural construct, part of a myth about global networks uniting the world in friction-free capitalism. But as this chapter has shown, convergence can, at times, "bite back" at the corporations promoting it. Some companies that seek to use convergence to control labor by reducing the workforce, deskilling workers, and employing temporary over full-time labor now find themselves facing converged trade unions like the CWA and CEP. By expanding their membership across the communication, information, and media sectors, these unions hope to improve their position when they confront convergent communication giants.

Convergent unions come in a variety of forms, ranging from the full-fledged convergence of the CWA and the CEP to the loose federation of unions exemplified by the National Alliance of Communications Unions that brings together the CEP and the TWU, plus one local of the CAW. Some unions, like the TWU, have pursued a limited form of convergence, expanding across provincial borders and into new media, but falling short of full convergence. And some, including the Hollywood unions SAG and AFTRA, have been left to figure out how to build a new working relationship after attempts at convergence failed.

As the case studies in this chapter show, labor convergence is neither easy to define nor easy to accomplish. And while unions see it is as a positive development, convergence does not in and of itself guarantee success at the bargaining table or on the picket line. In the case of the CBC, the solidarity achieved and maintained by the union was critical to its victory over management's effort to transform the workplace. But the fact that the employer was a Crown corporation, funded largely by tax dollars, ensured a significant degree of public interest in seeing the dispute settled. In addition, the locked-out workers included some of the most trusted and talented broadcasters in the country, people who were household names for Canadians. The union was able to take advantage of the public profile of its membership and its employer. Public support expressed for the strikers, at rallies on Parliament Hill and elsewhere, helped ensure that the union held together.

None of this applied in the case of the TWU. The workers were technicians, office staff, and installers—ordinary people, not household names. The employer was a telecommunications common carrier, not a national public broadcaster. Telus provides an important and federally regulated public service, but few Canadians feel the same way about their phone company as they do about the CBC. As a result, the locked-out Telus workers lacked the cultural capital available to the locked-out CBC workers. In addition, convergence does not come easily because it requires trust and a generally shared view of the state of labor relations, something the TWU had to face when a more conservative workforce from Alberta, and an unfamiliar workforce from Quebec and Ontario, joined the union. For the TWU, the social, political, cultural, and geographical obstacles to genuine integration were substantial.

As the failed merger between SAG and AFTRA showed in 2003, achieving unity is no easy task, even when unions share common members and common employers. Unions are political institutions, with their own governing and dues structures, bureaucracies, histories and, especially, cultures. For a significant number of SAG members, this includes a strand of craft-centered toughness, based on the belief that an exclusive union, limited to actors, would be a stronger union than one that includes other performers. As we showed in chapter 4, this appeal to craft can be effective in protecting a position, especially in the short term. However, by rejecting AFTRA—twice in less than a decade—the Screen Actors Guild has lost an opportunity to solidify its relationship with its closest ally. As a result, two unions that are each other's best colleagues find themselves having to devote resources to competing with each other, all the while knowing that those resources might better be spent promoting their joint interests. The SAG members who opposed the merger—a minority but a large enough one to determine the outcome of the vote in 2003—may see this as a small price to pay for retaining political and cultural independence. The current leadership of SAG believes its members will be better off because it is taking a tougher political stance against the industry and those in Washington who support the industry. Moreover, the status quo preserves and promotes the independent guild culture of SAG, a culture deeply rooted in the history of a labor organization that has struggled with seeing itself as a trade union. All this questions the perceived commonalities among communication workers that purportedly make them easier to organize and mobilize.

Labor remains a blind spot in communication policy research and that is unfortunate because labor issues play an increasingly important role in the policy environment. CBC workers argued that the public service mandate, which has largely been defined in terms of content produced for audiences, needs to take into account the nature of labor relations at the company because labor determines the extent to which the public service mandate can be met. The union maintained that only a company with the vast majority of workers in full-time, secure jobs, employing people with extensive experience in the business, can adequately serve the public as a national broadcaster. CBC management insisted that it could only carry out that mandate if it had the freedom to flexibly deploy contract workers throughout the corporation. Similarly, Telus argued that it needed flexibility, including the ability to contract out and outsource work, to effectively compete with Bell Canada and other firms in the converging electronic information services market. The union argued, among other things, that a nationally regulated telecommunications provider should provide full-time secure jobs to its workers. Is there an employment standard that should be part of the mandate for a nationally licensed broadcaster or telecommunications provider? Whatever one's position on the matter, there is little doubt that the analysis of labor issues would enrich the understanding and practice of communication policy.

In conclusion, the social impact of convergence, in its many manifestations, is an open question. That scholars in communication, sociology, and policy studies need to address it more systematically is not.

Chapter Six

Beyond Business as Usual: Social Movement Unionism and Information Workers

Trade unions occupy a unique position on the social landscape of North America. Their raison d'être is to give agency to workers collectively, which translates into individual gains as well. Unions have improved the economic, social, cultural, and political conditions of working people and, by extension, their families and their communities. At the same time, organized labor quite rightly claims responsibility for getting items like workplace health and safety laws onto the public agenda and eventually onto the statute books. In an information-based economy that makes it increasingly easy for companies to shed jobs and ship work abroad, labor unions are among a very small number of organizations dedicated to fighting the exploitation of workers on a global level. Given their significance, therefore, it is somewhat surprising that the decline of trade unionism in North America has been greeted by scholars and policymakers with indifference, and in some cases even with pleasure.

In the last few decades, however, some North American labor scholars have become increasingly concerned with trying to understand what the decline in trade unionism means, and what prospects remain for labor's renewal. Much of this work is prescriptive, based on the idea that given the importance of unions to the larger polity, ways must be found to protect and promote trade unionism. Some of it is descriptive, based on the empirical study of developments within bargaining units at the local level, between local unions and their communities, within individual national or international unions, and at the federation level. As the previous chapters have shown, there are signs of trade union renewal and revitalization at all these levels, and as the next chapter shows, this is occurring at the international level as well. However, the picture is complex, and the literature on trade union renewal reflects that complexity. (See for example Brody 1993; Bronfenbrenner, Seeber, and Oswald 1998; Nissen 2002, 2003; Raines, Bersen, and Gracie 1982; Tait 2005; Turner and Cornfield 2007; Turner, Katz, and Hurd 2001.)

It is possible to identify two key schools of thought in the scholarly literature on how to reinvigorate the trade union movement in North America. The first, which tends to be popular among industrial relations scholars, is known as "value-added unionism" or "mutual gains unionism." It urges unions to reinvent themselves as organizations that offer positive gains to employers, thereby making themselves more valuable in the workplace. The second, which has attracted the attention of progressive scholars and activists, is known as "social movement unionism." It seeks to reinvigorate trade unionism from the grass roots up, by overthrowing moribund bureaucracies and connecting unions with other worker organizations and community groups. These are not mutually exclusive ideas: value-added unionism and social movement unionism represent different points on a spectrum of approaches to solving the same problem, and one form does not necessarily preclude the other. This chapter begins by analyzing the problems with the "business unionism" model in North American labor relations, which both schools of thought identify, though in different ways, as a causal factor behind organized labor's decline. It then takes up value-added unionism and social movement unionism, assessing where and how they have contributed to rethinking the way North American trade unions operate. The chapter then looks at examples of new forms of organizing in the communication and information sector, forms that tend to adhere more to the social movement model than to value-added unionism. These include the Washington Alliance of Technology Workers (WashTech), which originated among contract workers at Microsoft, and a range of social movement organizations in Silicon Valley representing workers in the hardware production side of the computer business. It concludes by pointing out further prospects for social movement unionism within the communication sector.

Business Unionism and the New Economy

Though there are significant differences in interpretation, the contemporary visions of a reinvigorated labor movement emerge out of a generally shared conception of the reasons for the decline in trade unionism. These include the broader economic forces that form the environment within which unions operate, such as the shift from the factories of the industrial era to the information and service sectors that characterize postindustrial society, the rise of precarious employment, the emphasis on flexibility, the pressures of deskilling, the rise of neo-liberalism, and so on. They also include the "business unionism" model of industrial relations that has dominated labor relations in the

United States, and to a slightly lesser degree in Canada, since before the Second World War. Under this model, a union's dealings with an employer are based almost entirely on economic power relationships. The union serves as the bargaining agent for its members, negotiating and enforcing contracts, and in some cases administering pension plans or other benefit schemes. Business unionism does not preclude strikes; indeed, the ability to withhold labor is critical to asserting the union's economic power. But it tends to rely on membership compliance with a set of bureaucratic rules rather than on democratic decision-making, and it tends to encourage a focus on the workplace rather than on social activism or participation in the broader community. As a result, business unionism leaves unions vulnerable to the accusation that that they are outsiders to the employment relationship—third parties at the bargaining table rather than genuine representatives of the workers on the ground (Nissen 2003). Finally, business unionism tends to be less imaginative in organizing new or nonstandard members, or in envisioning how to reposition a union to cope with technological and social change.

While business unionism focuses on economic power and on how to arrange the best deal for the majority of workers in a bargaining unit, critics argue that money is only part of the employment relationship. Stone (2004) notes that organizational theorists use the notion of a psychological contract to characterize the nature of the expectations and perceptions workers and managers bring to the workplace. In other words, both see the employment relationship as one of reciprocal obligation, expressed through the creation of an internal labor market. In the internal labor market that has dominated North American labor relations since the middle of the twentieth century, worker expectations include the promise of long-term employment and a seniority system in which job security, pay, and benefits all increase with length of employment. In addition, workers expect firm-specific training and promotion through the ranks within that firm. They begin their careers at a low salary and at a low level of productivity, but as the years pass they become more productive and their wages rise. They are at their most productive in the middle of their career, but they earn their highest incomes at the end of their career. In the process, workers develop loyalty and commitment to the firm. They stay at the organization during their most productive years, believing that the employer feels a long-term commitment to them as well. The employer benefits, meanwhile, from the experience and stability of the workforce, and from the fact that the most productive workers are content to earn less than the most senior workers—and often less than they could if they moved to a different firm—because they believe that they too will earn higher salaries later in their careers. Stone notes that unions play an active role in constructing and policing internal labor markets. They do so by negotiating pay, benefits, grievance procedures, seniority systems, and narrow job classifications.

Business unionism, therefore, helps form the framework within which the psychological contract operates. In many ways, business unionism and the internal labor market described by Stone embody the so-called postwar social accord. At its simplest, the accord meant that labor agreed to reject or expel its more radical elements, especially communists, and to participate in capitalist society rather than try to overturn it. In return, capital agreed to provide workers with long-term security and a living wage. The state would serve as a referee between capital and labor and as a guarantor of benefits for both, through things like government-funded pension programs, taxation policies, and regulatory schemes that would remove irritants between workers and employers. As long as the economy continued to expand, capital, labor, and the state would all reap benefits from having a stable workforce in which a significant minority of workers—at its peak, about 35 percent—belonged to unions. There is considerable debate about the extent to which the ideal of this so-called Fordist workplace was put into practice (Foster 1988). But there is little argument that things began to change in the 1970s with erosion of the informal contract and the employment practices that went along with it. In some cases this was the result of competitive pressure on employers and on unionized workers. New technology offered employers a chance to deskill large elements of their workforce and shed jobs—especially jobs held by unionized workers. Corporate consolidation and concentration allowed for massive economies of scale. Employers in North America and other parts of the developed world began to abandon fixed mass production, moving the work to locales with lower labor costs—and usually, no unions. At home, managers began to develop a new model of labor relations, known as human resources management, which was predicated on marginalizing unions. Finally, the rise of neoliberalism in the 1980s emboldened employers to accelerate their attacks on trade unions.

All these factors promoted changes in the employment relationship and in the internal labor market, often in the name of flexibility and versatility. In the flexible workplace, the traditional features of the internal labor market—a seniority system, fixed job classifications, clear work rules, fixed hourly wages with the promise of higher compensation next year, and so on—came to be seen increasingly as drains on the employer rather than benefits to the employer. Stone writes that until the mid-1980s, employers tended to maintain a stable core in their internal labor market, but to supplement it with a flexible, peripheral workforce of subcontractors, temporary workers, on-call workers, and workers on other nonstandard working arrangements. After that, they began dismantling the core and repudiating the implicit contract for lifetime employment (Stone 2004, 70). Long-term workers were laid off through downsizing or restructuring. Some were rehired on a temporary basis. Others were told they should view themselves as free agents, not as long-term employees. New "regular" employees were told explicitly that they should not assume a long-term employment relationship. "Work has become contingent, not merely in the sense that it is formally defined as short-term or episodic, but in the sense that the attachment between the firm and the worker has been weakened" (2004, 70–71).

The psychological contract between employers and workers has also been rewritten. In place of job security and firm-specific training, employers offer "employability." In place of promotion opportunities, workers can expect "networking opportunities." In place of pay rates and benefits tied to seniority and longevity, the new psychological contract features "market-based pay." In place of collective bargaining and grievance arbitration, it offers dispute-resolution procedures for individual problems. For many workers, the price of this new psychological contract is precarity. Cranford, Vosko, and Zukewich (2003) note that one in three Canadian workers are employed in so-called non-standard arrangements, such as part-time work, self-employment, contract work, temporary work, on-call work, homework, telecommuting, etc.—a proportion so large that it raises the possibility that "nonstandard" is becoming the new standard.

Many of the core practices of unions are fundamentally at odds with this new work regime, Stone writes, though unions themselves are not necessarily incompatible with the new internal labor market and the rewritten psychological contract. She and other labor relations scholars contend, however, that if unions are to adapt to the new economy, they need to rethink the ways they operate.

Adding Value to Union-Management Relations

One approach, the "value-added" or "mutual gains" school of thought, challenges unions to dramatically alter their relationship with employers with an eye to developing partnerships that benefit both workers and management. In this view, "unions must reinvent themselves as value-adding organizations or networks offering positive gains to employers" (Nissen 2003, 134). In practical terms, this means they should avoid adversarial relations. They should also do away with some of the classic features of union-management relations in the business unionism model, such as seniority systems, fixed pay rates, or fixed job classifications. Because employers see these features as drains on their enterprises, unions should emphasize instead, and capitalize on, the values they bring to the workplace, such as their ability to upgrade the skills of their members, or their deep understanding of production processes and how to improve them. They should seek to create a "win-win" situation in which the employer gains the desired flexibility and cost efficiencies, and the union retains and extends its ability to represent worker interests. Advocates of value-added unionism believe that unions need new kinds of leaders, with new kinds of skills and talents. "Union leaders must develop new skills used in business management as formal bargaining and contract enforcement become less important. Problem-solving skills, rather than orchestration of adversarial conflict, become paramount" (Nissen 2003, 136). Some scholars note that unions must also revise their culture. In the past, union leaders who worked too closely with management were seen as sellouts. That attitude has to change, advocates of value-added unionism contend. In addition, local unions operating at the enterprise level should not have their hands tied by pressure from the national level to maintain industry-wide wage levels. As power shifts to the local enterprise level, national unions must reorient their goals.

Nissen notes that there have been numerous experiments in value-added unionism. Unions that have participated include the Union of Needletrades, Industrial and Textile Employees (UNITE), the United Auto Workers (UAW), the United Steelworkers of America, the Communications Workers of America (CWA), the International Association of Machinists, and numerous public-sector unions working in government (Nissen 2003, 136–37). Perhaps the most prominent and most thoroughly analyzed experiment was the partnership between General Motors and the UAW at the Saturn plant in Spring Hill, Tennessee, which lasted for close to twenty years. Conceived as an exercise in union-management partnership that might help address the challenge of competition from Japanese models, the Saturn experiment began in the early 1980s, with the union playing a role in designing the production process before the plant was even built. Once the plant opened, the union had input at all levels of decision making in the plant. All levels of management teams were staffed with both union members and managers. The experiment gave the union more power to run the workplace than in any other plant of its type (Rubenstein and Kochan 2001). But it brought with it a range of tensions. Because the contract was unlike those at other automotive plants, there were conflicts between the local union leaders in Tennessee and the national leaders of the UAW. There were also tensions within the corporation, which showed up in conflicts between Saturn management and GM management. For a time, Saturn competed well against Honda and Toyota. But sales peaked in 1994 and dropped 25 percent a year later (Peters and Maynard 2005). GM resisted extending the Saturn model to other plants and in the end, the union-management partnership was dissolved. In 2004, the union agreed to a contract that ended Saturn's special labor agreement (Bernard 2004). The plant continues to be in turmoil. Recently, GM secured an agreement on flexible work rules at Spring

Hill after promising workers to build a new Chevrolet crossover SUV there (Terlep 2007). It would be simplistic to see the cancellation of the labor-management partnership at Saturn as simply a failure of value-added unionism. Rather, it reflects difficulties in the North American car manufacturing business, and especially the strong competition from automakers like Toyota. Nonetheless, the Spring Hill story shows that an employer facing severe competitive pressure responds by trying to extract more value from its workforce. In other words, the pursuit of profit, which is capitalism's prime directive, takes precedence over the pursuit of partnerships.

In the communications sector, one of the pioneers in the value-added approach was the CWA. Former president Glenn Watts advocated cooperative labor relations and strategic partnerships with management, and pursued one such partnership with AT&T. It fell apart, however, when AT&T repeatedly made business decisions that undermined the union (Nissen 2003, 143–44). Some printing trades unions have also practiced labor-management cooperation by helping employers to manage their jobs efficiently.

While some attempts at value-added unionism have been successful, the idea has fundamental weaknesses. One is the suggestion that by taking a more cooperative stance toward each other, labor and capital will be able to overcome their historical and systemic antagonisms. Value-added unionism is predicated on the notion that labor and capital both can and will embrace the idea of partnership. However, there is an implied expectation that in creating a partnership to overcome those systemic or long-standing antagonisms, the "partner" that needs to make the most significant changes is labor. The term, after all, is "value-added *unionism*," not "value-added *management*."

In addition, Nissen notes, critics of value-added unionism say it raises a fundamental question: why do workers join unions in the first place? Is it to help their employer? Not likely. Instead, workers believe collective action is the best way to obtain industrial justice. Therefore, the challenge for the proponents of value-added unionism is to convince workers that industrial justice can still be attained under an arrangement whose purpose seems to be based on helping employers in their drive for higher profitability and economic competitiveness. What does this mean for workers? If unions give up their traditional sources of power in the workplace, will they have sufficient power to force the employer to accept union-friendly policies? As Brody puts it (1993, 263), giving up outdated work rules might make strategic sense; "retreating from 'them and us' as a basic orientation is a different matter."

Value-added unionism may face its best prospects in situations where the union already has a strong position in relation to the employer, as was the case at the Saturn plant in Spring Hill, or in situations where the union and the employer occupy a dominant and near-monopoly position in the industry, as was the case at AT&T. In such cases, the employer may see clear benefits in working in partnership with the union, and these partnerships may last for fairly significant lengths of time.

But this raises another question: How applicable is the value-added approach in organizing new workers? The answer is mixed. Employers that are already "worker friendly" are much less likely to be the target of organizing campaigns than employers that are seen by their workers as greedy or exploitative. Workers in the former kind of enterprise may already belong to a union, or they may have little or no interest in joining a union. Organizing occurs most often out of a sense of injustice, out of a feeling that management is not cooperative, or caring, or concerned with the needs of workers. Especially in the communications area, the injustice is not always simply economic. Organizing drives among journalists at Canada's *Calgary Herald* and *London Free Press* newspapers, among others, occurred when journalists felt management changes threatened the quality of their journalism (McKercher 2002; Nagy 2007). The workers sought union certification in part to tip control over the journalism away from corporate managers and toward workers who felt a deep connection with the community. Nonetheless, the growing ranks of knowledge workers, especially the creative or highly educated information workers who see themselves as professionals, may find some appeal in joining a union that is comfortable speaking the language of productivity and efficiency. These workers might be more inclined to see advantages for workers in flexible work regimes, or market-based pay, or the promise of networking rather than job security, and they might be willing to participate in a partnership with management. However, that still leaves the question of what incentive management has for welcoming such a union. Not much, according to Brody: "No amount of union enthusiasm for cooperative relations and employee involvement is likely to persuade employers that collective bargaining is preferable to a union-free environment" (1993, 258). Ultimately, then, the challenge for proponents of value-added unionism is to figure out ways to give employers the flexibility they desire, but to retain or restore sufficient worker power to make the struggle to organize worthwhile.

Social Movement Unionism

While industrial relations experts have focused on altering labor-management relationships in ways that add value, others with an interest in labor—such as progressive intellectuals, labor educators, and community activists—propose a different approach. The school of thought known as social movement unionism also begins with a critique of business unionism and calls for unions to reinvent themselves. But its diagnosis of the problem created by business unionism is different, and so is its proposed remedy.

Social movement union theorists argue that especially in the years after the Second World War, trade unions became servicing bureaucracies, dominated by their leaders and staff. As relations with employers became routinized and dominated by the legalities of negotiating and enforcing a contract, the rank-and-file character of labor was lost. Within the unions themselves, meanwhile, business values began to dominate the way the organizations dealt with their members. Members came to see their unions as institutions or businesses that provided a service to their customers—in other words, the members. These customers paid a fee for this service through their union dues. One critic put it this way: business unionism encouraged members to see their union "as a vending machine rather than a social movement of solidarity and direct action" (Gibson 2007). Some social movement union theorists argue that as a result the labor movement was, in fact, no longer a movement at all.

Nissen (2003) notes that social movement theorists argue that business unionism never worked well for organizing the unorganized, but its weaknesses were hidden in the three decades between 1945 and 1975. During these years, relatively high union density meant that unions had genuine influence in the workplace and were able to ensure that their members shared in the wealth generated by postwar economic expansion. Unions also had political clout. In the United States, this was centered in the national labor federations, which united in the 1950s to form the AFL-CIO. In Canada, unions exerted political influence through similar national labor congresses and through the labor-affiliated New Democratic Party, which could typically count on the support of between 10 and 20 percent of the electorate nationally.

But as the industrial economy began to break down, the weaknesses inherent in business unionism were laid bare. Although there were exceptions, unions were unable to rally their members to respond to assaults from their employers, or to mobilize effectively even within the workplace. As Nissen puts it, members trained to view the union as a servicing bureaucracy were not inclined to participate, "any more than they wanted to participate in an insurance company from which they bought a policy" (2003, 140). Unions were equally unable to connect with people in the broader community. Public approval ratings of unions dropped. Again with exceptions, much of organized labor was isolated from, and often hostile to, the social movements of the 1960s, such as the civil rights movement, the women's movement, the antiwar movement, and the environmental movement. As a consequence, women, young people, and people of color—the very people organized labor needed to engage—saw organized labor as old fashioned and out of touch.

The social movement school of thought argues that if organized labor is to survive, it must return to its rank-and-file roots. This means building a social movement of workers and their allies to combat capital's domination of every facet of daily life. In this view, unions should become champions of those oppressed by the U.S. economic system, as they were in the 1930s. They should make every effort to organize the unorganized, including groups traditionally seen as unorganizable, such as contingent workers, recent immigrants, women, and minorities. They should also engage more vigorously at the political level, regaining a place on the national stage (Gupta 2006; Turner and Hurd 2001).

Robinson (2002) notes that the concept of social movement unionism first gained popularity with scholars of labor movements outside of North America, in countries like South Africa, Brazil, South Korea, and the Philippines, where trade unionism fed into and grew out of wider movements that pushed for social change. More recently, activists and scholars have begun applying the term in North America, as both a characterization of some unions and as an ideal toward which organized labor should aspire. The picture that is emerging of social movement unions in the North American context generally focuses on a set of characteristics that, taken together, distinguish these unions from those practicing business unionism.

In offering a detailed definition of social movement unionism, Robinson describes an "ideal type" for North America (2002, 190–93). He writes that social movement unions tend to have a high degree of autonomy from the state, political parties, and employers. Membership is largely voluntary. The unions tend to be very democratic in their internal political processes, which means they are accountable and responsive to their membership. While leaders lead, they do so in close interaction with the members. Because members do not have to belong to the union to keep their jobs, they believe their leaders are accountable to them. In addition, members feel a strong sense of ownership of their union. This results in an unusually high capacity to mobilize members. Social movement unions are generally not focused on a single issue, or simply on workplace issues. Rather, they believe that their union has a mission to bring about social change. The changes sought are fundamental adjustments in the political, social, or economic realm. For instance, Robinson notes that social movement unions in South Africa or Brazil have fought for things like the abolition of the colonial subordination of a nation, the democratization of the political order, the radical transformation of the capitalist mode of production, and the end of apartheid and other forms of oppression. In this sense, he notes, social movement unions have a radical culture. Social movement unions typically see themselves as highly inclusive, in terms of whose interests they represent (e.g., all members of a subordinated nation, or all workers) and whom they seek to organize. Because the economic and political elite are unlikely to want to support groups that offer such fundamental challenges to the social order, social movement unions tend to seek allies among broader social movements that share some of their social change goals. Community organizing is an important supplement to workplace-based organizing—and sometimes, especially in the early days of a social movement union, is a substitute for it. These unions may have allies in political parties, but they seldom rely solely on electoral politics as the way to promote or fulfill their goals of long-term social change. The goals of a social movement union are shaped by the union's culture and the environment in which the union operates. The former includes the union's moral economy (how critical it is of the status quo) and its collective identity (how inclusive it is in terms of the interests it represents). The latter includes a series of conditions that either favor or discourage the formation of social movement unions, key among them worker collective identities, state policies, and the state of the economy.

Nissen writes that social movement unions find moral legitimacy in their ability to be instruments for economic and social justice. Their appeal is to working-class people in general, as well as to people in coalitions or movements that fight against corporate domination of daily life. Corporate greed typically provides a common target for organized labor and social movements (2003, 145–46).

None of the worker movements in the United States or Canada displays the full set of characteristics in Robinson's "ideal type" of social movement unionism, though as we will show below, several exhibit one or more of these characteristics. Nonetheless, a number of scholars have found a considerable increase in union social movement activism in North America, starting in the 1990s. The intensity and orientation of this activism varies considerably, as do the players. Some emphasize joint efforts by labor and social or community movements to press for social change, including a better deal for workers. Some are focused more narrowly on organizing or on contract campaigns that attempt to build public support for labor. And some, especially among communication, media, and information unions, result in the creation of non-traditional labor organizations to represent the needs of workers who, for any number of reasons, are unwilling or unable to join traditional unions.

In community-based or contract campaigns, central labor councils have become increasingly significant actors. These groups bring together union locals operating at the community level and are the oldest labor federations in several parts of the United States and Canada. Central labor councils have been key players, for example, in the host of living wage campaigns that have spread across the United States, especially since the 1990s. These campaigns generally focus on passing local or county ordinances to raise the pay of poor workers. The driving idea behind these ordinances is that public money shouldn't be used to create or subsidize a workforce doomed to sub-poverty-level incomes. Participants in a campaign vary according to the locale, but generally include labor bodies, churches or religious-based social justice organizations, and poverty or low-income organizing groups. The ordinances themselves also vary. Some focus on employees of the city itself, others cover employees of service contractors that sign contracts with the city, and some include companies that receive public subsidies. (See, for example, Levi, Olson, and Steinman 2003; Nissen 2000; Reynolds and Kern 2001.)

Labor councils have also become increasingly active in supporting or promoting union organizing campaigns, especially councils whose leaders have what Gapasin and Wial (1998) call a transformative orientation. By this they mean leaders who see central labor councils as vehicles for social change, and as organizations that can bring affiliated unions together with each other and with nonlabor community organizers to advance the interests of the workers and the community. Gapasin and Wial describe one campaign in which the labor council worked with an environmental group to support unionization of an employer that was a major polluter. At other labor councils, leaders strategize frequently with leaders of community organizations and with individual unions, acting as nodes in a broader network of socially progressive organizations that are concerned with work and employment issues. Labor-community coalitions have fought against plant closings, and fought in campaigns for corporate accountability. In some cases, these coalitions play a decisive role in union contract campaigns, which are aimed at rallying members and building community support to help a union win a new contract. (See Craypo and Nissen 1993; Haines and Klein 1982; Heckscher and Palmer 1993; Nissen 2004; Peters and Merrill 1998; Peterson 2002; Turner 2004.)

Social movement union activism may exist at the local, regional, or national level. For example, the AFL-CIO's Jobs with Justice campaign relies on activists in unions all over the country. Founded in 1987, the campaign aims to create a national coalition to fight for workers' rights based on three demands: employment security, a decent standard of living, and the right to organize on the job. To achieve these goals, it seeks to build powerful labor-community coalitions. At the local level, these coalitions push for governments and corporations to behave responsibly in their communities, and show unorganized workers that collective action can be effective. At the state and national level, the goal is to push for social and legislative changes favorable to workers. Partners in Jobs with Justice include nonlabor groups such as the National Organization of Women and the Southern Christian Leadership Conference, a major civil rights organization.

Individual unions have worked with community groups, including religious, antipoverty, or environmental groups, to organize workers or to press for contracts. Nissen (2003) identifies several that show the characteristics of social movement unionism, including the United Mineworkers, United Electrical Workers, and the Communications Workers of America (CWA), a union that, as mentioned earlier, has also experimented with value-added unionism. Nissen points out that the CWA has a high capacity for mobilization. It routinely runs contract campaigns, has built alliances internationally, and has used its own members to organize the unorganized. In addition, its current president, Larry Cohen, was one of the activists who founded Jobs with Justice (Nissen 2003, 141–44).

However, the best-known example of social movement unionism is the Service Employees International Union (SEIU). With 1.9 million members, the SEIU has grown from being a substantial but relatively unknown union—it had 625,000 members in 1980—to become one of the largest and fastest-growing unions in North America. SEIU grew at a time when other unions were shrinking. It did so, in large part, by setting out to reinvent itself as a social movement union. For example, it hired activists with experience in social movements, launched campaigns to organize service workers of all types, and ended up with a membership that is among the most diverse in the United States. Its enormously successful Justice for Janitors campaign, which began in 1985, has targeted people who are traditionally considered to be difficult to organize and therefore rank low on the priority list of other unions: non-English-speaking immigrants and African Americans who earn very low wages. It has also successfully organized workers in hospitals, nursing homes, and home health care, many of them immigrant women and minorities. Many of the union's successes have been due to the fact the SEIU works with community groups and religious and political leaders on a range of levels. A Florida campaign to organize nursing home workers, for example, drew on interventions from the local Jobs with Justice chapter, the NAACP, local Haitian organizations, the local interfaith committee for worker justice, and progressive local African American and Hispanic politicians (Nissen 2003, 142). In 2005, SEIU and the United Food and Commercial Workers launched separate media campaigns against Wal-Mart, the largest retailer in the United States and a notoriously antiunion company. Both campaigns sought not necessarily to organize workers, but to create a public movement to reform the company's business practices, particularly on wages and benefits (Fones-Wolf and Fones-Wolf 2007).

The SEIU's tactics and organizing successes have contributed to the growing interest in the phenomenon of social movement unionism. Nonetheless, the idea raises some difficult questions. One has to do with whether North American trade unions are really attuned to, or ready to engage in, activism on a broader stage. Kimeldorf has argued that regardless of their political orientation, American labor unions have persistently engaged in a form of syndicalism that is focused on the workplace: "a kind of 'syndicalism pure and simple'—defined by its point-of-production focus, aggressive job control and militant direct action" (1999, 15). In the American Federation of Labor, this appeared as a kind of business syndicalism. In the Industrial Workers of the World, it was a more radical industrial syndicalism. But in both cases, he argues, the focus was on righting workplace wrongs. This suggests that while labor unions can be a part of broader social movements on civil rights, the environment, or globalization, their members are mainly concerned with how these issues play out in the workplace. Meanwhile, unions that attempt to shift resources to external organizing can create internal turmoil (Fletcher and Hurd 1998). There may be disputes over appropriate ways to spend resources and questions about whether members' dues—under business unionism, the fee they pay for services—should be spent on social movements that are only indirectly related to the workplace. Social movement union theorists argue that rank-and-file members need to be convinced that the union should be more than a servicing agent, and they need to be genuinely interested in developing membership control of the organization. But this requires a substantial change in attitude among both members and union leaders.

Some authors have argued that unions work most successfully in community coalitions in cases where the coalition is focused on labor issues. In other words, they thrive in narrow, vanguardist coalitions where the other participants support labor's agenda and see their own interests as secondary. But unions are unable to take a backseat in multilateral coalitions that work for broader social goals, including but not limited to labor. Nissen's study of labor-community activism in South Florida (2004) questions this view and concludes that it is indeed possible to create common cause. Doing so successfully, however, depends on many different factors, especially the presence of individuals who are willing and able to build bridges between unions and the community.

Social movement unionism's biggest potential appears to lie in organizing and mobilizing, and in creating or contributing to new forms of worker organizations that organize and mobilize but do not necessarily engage in traditional collective bargaining. Because it emphasizes social, economic, and political justice for workers, social movement unionism would appear to have much more potential legitimacy among the unorganized than value-added unionism. Workers might be more likely to join an organization they feel is fighting for their rights as workers, not as supports to capital. This is even more the case when the organization has the support of interfaith groups, social justice organizations, or other community groups. Social movement unionism also seems to hold some potential to give workers the strength to fight back against the economic ravages of neoliberalism and globalization, and to build bridges between labor movements in the developing and developed world.

Nissen notes, however, that rank-and-file union members in the United States are not demanding, in any great numbers, that their unions take a wide-ranging role in insurgent social movements (2003, 148). In Canada, where the union movement has traditionally been closer to social movement union ideals than in the United States, unions have also struggled with the idea. Indeed, Robinson (2002, 215–220) contends that interunion competition has increased in Canada, and disagreements are growing over basic political strategy on whether, and how strongly, to support the New Democratic Party. The most intense conflicts occur when workers who belong to one union decide to switch to another. Some scholars point out that much of this switching relates to continuing efforts to create genuinely Canadian unions in place of so-called international unions, which are in effect U.S. unions with a small Canadian membership. The proportion of Canadian trade unionists belonging to international unions declined from more than 70 percent in the mid-1960s to less than 30 percent in the mid-1990s (Dawes 1987; Murray 1995). This occurred through the creation of new Canadian unions as well as through the secession of Canadian branches of international unions. But as a result, unions that should be allies in social movements, such as the SEIU and the Canadian Auto Workers, for example, find themselves trading accusations over "raiding" each other's members. Nevertheless, several Canadian unions, including the CEP, have been in the forefront of campaigns for workplace justice, and social and political reforms, both in Canada and abroad.

Worker Movements in the Information and Communication Sector

One sector of the economy that has seen wide-ranging experiments in social movement unionism is the communications and information sector. These efforts have ranged across the class spectrum, from trying to organize highly skilled, well-paid high-tech workers in Microsoft's backyard to working with low-paid, immigrant pieceworkers, most of them women, who put together cables and circuit boards in their homes in Silicon Valley. In recent years, we have also seen growing solidarity efforts among the so-called precariat of information workers, including freelance writers, editors, and software designers, who have created associations that are neither typical unions nor traditional professional groups but rather a combination of the two. Many of these new information workers' organizations are allied with other groups in the community and display features Robinson would see as characteristic of social movement unions, including open memberships and a social agenda that seeks to bring about changes in the social, economic, and political realm. In this section, we examine some of those efforts.

In late 1997, a group of high-tech workers in Redmond, Washington, ran out of patience with their work arrangement. They were working full-time at Microsoft—accounting for 30 percent of the workforce—but were classified as temporary workers. This meant they lacked the status, perks, and benefits of their colleagues on the permanent staff. They received "no health insurance, no sick leave, no vacation pay, no pension plan, no 401(k) program, no stock options and absolutely no job security," according to Marcus Courtney, who had worked as a Microsoft temp for three years (Deibel 1999). Indeed, these temporary workers didn't actually work for Microsoft itself. Many had been interviewed and hired by Microsoft managers, but were told to sign on with one of several temporary agencies in the area. Months or even years later, they were still stuck with the agencies. The workers were told they couldn't switch agencies voluntarily, but could be assigned to a different agency. At Microsoft, the contractors were treated differently than permanent workers, even when doing the same work. They had to wear colored nametags identifying them as temps. They were ineligible to play on company teams, shop at the company store, or participate in launch parties to celebrate the release of new products. They felt exploited and excluded (Brophy 2006; van Jaarsveld 2004).

The contractors—or "permatemps," as they came to be known—had been unhappy with aspects of their work for some time. In the early 1990s, they filed a class action lawsuit seeking the same kind of benefits as permanent workers. They would eventually win a partial victory in that case, and would begin collecting $97 million from Microsoft in 2005. But in late 1997, the Washington State trade alliance that includes Microsoft and other high-tech employees asked the state labor department to adopt a rule change that allowed employers to eliminate time-and-a-half overtime pay for workers who earned more than $27 an hour. When the state agreed in early 1998, the permatemps hit the roof. They saw this as a clear instance of workplace injustice, perpetrated by a state agency that took its cues from industry. They also believed it was anything but an isolated case. "The industry can ask for something else six months from now and they'll probably get it because there is no countervailing voice, no workers' representation," technical editor and permatemp Mike Blain told the *Seattle Times* (Fryer 1998).

Courtney said he and Blain had already been talking with each other about labor issues in the high-tech area, having been introduced by an organizer at the King County Labor Council. They came up with a plan to launch a website aimed at contract workers. But when the state agency adopted the change on overtime pay, "there was this big spontaneous outburst and outpouring of support against the ruling by tech workers" (Interview with E. Brophy 2005). He, Blain, and other discontented contractors used the momentum to form the

Coordinating Committee for High-Tech Labor Issues, which Courtney described as a group of union and tech workers exploring the possibilities of what organizing in the high-tech sector would look like in Seattle. "From that it got pared down basically to the National Writers Union and the CWA and the King County Labor Council and tech workers," Courtney said.

These groups had different interests, but all had some experience with the issues facing the temporary high-tech workers. The National Writers Union (NWU) is a professional group, but also a trade union for freelance and contract writers: journalists, book authors, business and technical writers, web content providers, and poets. It has roughly two thousand members in sixteen chapters across the United States and since 1992 has been affiliated as (UAW) the United Auto Workers. The union describes its mission as working to defend the rights and improve the economic and working conditions of writers, and sees its connection to the UAW as a way to recognize the common aspirations writers share with other working people, "and common needs for peace and justice, healthcare, education, a livelihood, and democratic institutions" (NWU 2007). As such, it serves as a bridge between traditional organized labor and independent knowledge workers. The CWA, as we have described in earlier chapters, is an example of labor convergence in action: a large trade union that brings together workers across the communication, media, and information sector. In recent years, it has become increasingly interested in organizing beyond its core of telecommunications workers. The King County Labor Council, meanwhile, saw promoting and assisting organizing as part of its mandate. It helped the contract workers set up six public meetings, held at the labor council's headquarters, to discuss issues, and it offered recruitment advice as the organization came together.

The fledgling group of temporary workers built up a phone bank and a committee of contract workers who wanted to move organizing forward. The result was the Washington Alliance of Technology Workers, or WashTech. In Courtney's words: "We had a CWA organizer who was with us from the very beginning . . . [and] she was very open to the idea of, let's just see how this plays out" (Interview with E. Brophy 2005). But he stressed that the idea of organizing didn't come from the CWA; it came from the permatemps. WashTech launched its own website in May 1998. A few months later, in August, its members agreed to accept an offer to affiliate with the CWA.

Right from the beginning, WashTech was an unconventional union, and although its membership and focus has changed significantly over the last ten years, it remains so today. It combines elements of traditional trade unionism with features that are characteristic of social movement unionism. Like the latter, it is democratic and grassroots oriented. Membership is open to any tech worker who wants to join. Nonmembers can subscribe, free of charge, to its listserv. The organization pursues an agenda that combines working on standard workplace issues, like health care and job security, with calls for social and economic change, especially on the issues of offshore outsourcing and exploitation of visa workers. Like traditional trade unions, it is focused on the needs of a particular group of craft workers, though its definition of craft has been broadened steadily. WashTech organizer Karen Estevenin said the group now refers to itself as an organization of "tech" workers, a term that is considerably more elastic than "high-tech" workers (Interview with E. Brophy 2005). It also seeks to organize bargaining units and to engage in collective bargaining. WashTech represents, therefore, an experiment in creating a new form of worker organization: a hybrid of traditional trade unionism, information clearinghouse, and social activism.

Much of this has to do with the composition of its original membership. WashTech was created to represent a type of worker that is extremely difficult to organize: the highly educated, well-paid if precariously employed, independent-minded young Microsoft worker. These people saw themselves as entrepreneurs and professionals, who chose to work in a field that held out the possibility of joining the ranks of the superrich. Although these workers had benefited from growing up in the relatively comfortable years of the postwar social accord, they came of age at a time when labor unions were in decline. Many had little knowledge of, or respect for, trade unionism. As a result, WashTech leaders, especially in the early years, avoided using any language that smacked of trade union jargon. They stayed away from words like "union," "labor," or "organizing." According to Blain, "some people balked at the word 'worker' because they thought it was too militant. They wanted to use 'employee' or 'professional'" (Rothenburger 1999).

Without a conventional base of dues-paying members, WashTech needed financial support. The CWA provided it, and it continues to subsidize the local today. It also needed help from experts in organized labor. The CWA assigned the task of working with high-tech workers to the Newspaper Guild sector of the CWA. The guild, a union of journalists and other white-collar news workers, had merged with CWA in the mid-1990s. The decision to put WashTech with the guild was made in part because of local conditions. In the Pacific Northwest, the guild local is a large one: it currently counts twelve hundred members in twenty different units, including at the *Seattle Times* and the *Post-Intelligencer*. But it also made strategic sense. High-tech workers weren't unique in resisting trade unions because they saw themselves as professionals: sixty years earlier, journalists had felt much the same way. Indeed, in 1933 the founder of the Newspaper Guild, Heywood Broun, mocked journalists for being "peculiarly susceptible to such soothing classifications as 'professionals,' 'journalists,' 'members of the fourth estate,' 'gentlemen of the press' and other terms which have completely entranced them by falsely dignifying and glorifying them and their work" (quoted in McKercher 2002, 83n6). And yet the Newspaper guild was able to organize and represent news workers who are as independent, iconoclastic, and difficult to manage as high-tech workers. The CWA saw the guild as its best link to professionals in the labor movement. It felt that high-tech workers would be more comfortable working with media workers in the guild sector than with the telecommunications workers who historically formed the core of the union. Finally, the decision to make the link between WashTech and the guild was seen as a good move for the guild, which was adjusting to belonging to a union comprised mainly of technicians or craft workers. The campaign to convince Guild members to merge with the CWA had been waged largely on the idea that labor convergence was an important step for communication and media workers (McKercher 2002). Working with WashTech would be a sign of labor convergence in action.

The permatemps worked at Microsoft, but not directly for Microsoft. This meant it would be virtually impossible to organize them into a conventional bargaining unit. All the agencies would have to be brought together simultaneously—clearly an impractical task. Just as problematic was the fact that many potential members did not buy into the idea that they needed a union. As van Jaarsveld notes (2004, 380), the "persistence of antiunion attitudes among potential members," as well as in the industry, was—and still is—one of the biggest challenges facing WashTech.

Instead of running a conventional organizing drive, WashTech decided to offer membership to anyone who wanted to join. It focused on running its listserv, acting as a clearinghouse for information on the plight of contractors and other technology workers, and pursuing a social and political agenda that concentrates on publicizing and combating the exploitation of a global high-tech labor force. As Courtney told a reporter: "We're trying to create a new model of unionism. In this industry, collective bargaining is not an attainable goal right now. That's not going to preclude us from organizing and building power for workers on the job, at the legislative level and in the community. If we play by the rules, we'll never win. So we're going to build a new kind of worker organization that works for this industry and new economy jobs" (Moberg 2000).

In early 1999, Microsoft took a series of steps that addressed many of the financial irritants that had contributed to the creation of WashTech. It announced that it would change its contracts with employment agencies to improve benefits for the permatemps. Instead of being told by Microsoft which payroll agency to work for, the contractors would be able to choose from among competing agencies. And all agencies were expected to provide at least thirteen paid days off a year, medical and dental coverage, training opportunities, and a retirement program. At the same time, however, the company was pursuing other policies that angered the contractors, including mandatory "breaks in service" between contracts—at first thirty-one days, and later extended to one hundred days. Most contractors took their break and returned to Microsoft on a new assignment (van Jaarsveld 2004, 372). If the companies and agencies thought this might derail WashTech, they were disappointed.

WashTech began a program of trying to organize specific groups of workers into conventional bargaining units, starting with Tax Saver, a highly specialized working group within Microsoft. At the time, labor law required the group to obtain consent from both Microsoft and the temporary employment agencies. The companies refused, but the group pressed on, raising questions about working conditions and attracting significant media attention. Shortly thereafter, Microsoft sold off the products produced by the Tax Saver group and about one hundred workers, including the activists, lost their jobs (van Jaarsveld 2004). An organizing campaign at a computer company in Frazer, Pennsylvania, ended in 1999 when the workers withdrew their petition to join WashTech after heavy lobbying from management (Rosa 1999). Workers at Amazon.com, having heard about WashTech, contacted the union in 1998. A few years later, the union launched a large-scale organizing drive among customer service workers who were based in Seattle. It ended in early 2001 when Amazon laid off the Seattle workers and moved the jobs to a customer service center in India (van Jaarsveld 2004). With the help of the CWA, WashTech successfully organized close to one thousand workers at Cingular wireless in 2005 and negotiated a three-year collective agreement a year later. According to its website (www.washtech.org/about), WashTech has signed a total of four collective bargaining agreements and is "in the process of signing even more."

But traditional organizing and collective bargaining are only part of WashTech's mandate. Much of its work is focused on what van Jaarsveld calls "mutual benefits activities" (2004, 364), such as information sharing among members and training services, and on political action on two key issues: limiting the importation of high-tech workers to the United States on H-1B visas, and curtailing offshore outsourcing of high-tech work. It sees both issues, like the "permatemp" question, as symptomatic of the industry's desire to expand the labor pool, reduce wages, and turn workers into commodities. Courtney put it this way: "The global economy makes no distinction whether you manufacture Fruit of the Loom underwear or you manufacture a high-end software product. The only difference is who can do it cheaper, and who can do it the fastest" (Interview with E. Brophy 2005).

The organization has collected and circulated information, and lobbied for rule changes and new legislation at the federal and state level. This has helped propel its membership from two hundred dues-paying members and fourteen hundred listserv members in 1999 to fifteen hundred dues-paying members and seventeen thousand listserv members in 2006 (Rodino-Colocino 2007).

Rodino-Colocino suggests that in fighting the expansion of the H-1B visa program, and the exploitation of the workers who hold such visas, WashTech made a transition from concentrating on the concerns of local workers to engaging with issues facing the global labor market. The H-1B visa program grants temporary work permits to technically skilled workers who fill jobs left vacant by a shortage of skilled U.S. workers, most often at reduced levels of pay and ever more precarious employment conditions. In theory, it allows employers to hire the most qualified candidates in the world. In practice, however, it drives down labor costs in the United States and depresses incomes for domestic high-tech workers. Rodino-Colocino (2007) writes that this occurs directly, because visa holders are often paid less than U.S. high-tech workers, and indirectly, because visa holders are less likely to fight for better pay, hours, or working conditions. These workers—Paula Chakravartty (2005) would call them the "weak winners of globalization"—are afraid that resisting exploitation by their employers could cost them their jobs, and in turn cost them their visas. WashTech has been fighting steadily against the expansion of the program vigorously sought by Microsoft and other high-tech companies.

Rodino-Colocino (2007) writes that WashTech's concerns with offshore outsourcing were sparked in 2000, around the time of the dotcom bust. Two specific incidents prompted WashTech to tackle the issue. The first was Amazon.com's decision to lay off customer service workers in Seattle, whom WashTech was trying to organize, and send the work to India. The second was the revelation in 2002 that Microsoft was planning to send "critical" (i.e., highly skilled, high-paying) work to India. The union obtained leaked plans, which suggested that sending work to India could cut costs in half. WashTech publicized Microsoft's plans by posting them on its website and alerting the media. To underscore offshore outsourcing's effect on high-tech job loss in the United States, WashTech and the nationally focused website, Techs Unite.org, launched the Offshore Tracker in 2004. The tracker combines reports of offshoring in the news media and invites individuals to file reports detailing such events. Upon verifying the event and the number of jobs lost, union staff post the results. At the end of 2007, the tracker reported that close to 530,000 jobs had been sent offshore since January 2000, and more than 260,000 jobs lost.

WashTech has also focused on mutual aid for tech workers. It provides access to discounted training courses for members and since 2001 has operated, along with the parent CWA, a training center for high-tech workers in Seattle. This center is also used for CWA's training partnership with Cisco Systems. Its website contains links to job postings, and to a benefit program for union members that offers discounts on a range of consumer products and services. The emphasis on worker aid extends to WashTech's lobbying efforts. In 2007, for example, it helped convince the U.S. House of Representatives to include service workers in a major overhaul of the Trade Act Adjustment Assistance program, which helps workers who lose their jobs as a result of increased imports from, or shifts in production to, foreign countries. Though the president vowed to veto the bill, the passage of legislation expanding the coverage to workers like software engineers and customer service employees was seen as a victory for high-tech workers (Courtney 2007a).

Van Jaarsveld writes that WashTech's strengths lie in its willingness to experiment with strategies, and to focus its efforts beyond simply providing services to members. Its approach is "rooted in mutual aid logic and political action" (2004, 381)—a phrase that resonates with the literature on social movement unions. It has also sought to make connections with other organizations that share its interests. The TechsUnite website, for example, is a project of the CWA and WashTech, but it draws on the support of a range of groups. These include the Center on Wisconsin Strategy (COWS), a national policy center that describes itself as a "field laboratory for high-road economic development" (COWS 2007); Working Today, a national nonprofit membership organization that promotes the interests of people who work independently; and Alliance@IBM, which like WashTech is affiliated with the CWA.

WashTech offers one model for social movement unionism among information workers: a model that grew out of the interests and needs of highly educated, precariously employed "creative" workers. As Pellow and Park (2002) describe, efforts at promoting labor rights in Silicon Valley have had a different starting point and serve a different group of information workers. Operating alongside the high-end software design facilities of Silicon Valley are low-wage and high-hazard work sites populated by poor people, immigrants (both illegal and legal), and people of color. In Silicon Valley, between 70 and 80 percent of the production workforce is composed of Asian and Latino immigrants, most of them women. The chip production industries and the companies that assemble printed circuit boards, cables, and printers are nonunionized, and the companies are determined to keep them that way: workers who try to organize are "humiliated, denounced, and fired as companies can neutralize opposition, downsize the workforce or relocate to another state or country" (2002, 2).

Pellow and Park show that workers employed in the so-called clean rooms, where chips are produced, have been sickened or killed by years of exposure to toxic chemicals. Assembly workers, who put together cables, circuit boards and printers, are in even more dire straits. Some work in factories but others work at home on a piece-rate basis. Regardless of the location, they often work for less than a living wage, on temporary contracts with few or no benefits, and with exposure to toxic substances and other hazardous conditions. These assembly workers are at the bottom of the hierarchy of companies that make chips and electronics products: "invisible, unorganized, fragmented, and isolated from each other" (2002, 167). Meanwhile, although Silicon Valley is home to communities that are models of cleanliness and environmentally sensitive design, the immigrant workers at the low end of the production ladder live in communities that have been polluted with waste from the electronics industry. The Mountain View community of Santa Clara County has ten federally designated toxic sites—more than anywhere else in the United States—and a second community, Sunnyvale, has six (2002, 81). Both are home to working-class communities of immigrants and people of color.

In Silicon Valley, the fight for labor rights is bound up with the fight for health and safety, both in and beyond the workplace. Pellow and Park see it as part of the environmental justice movement, which draws on links between the workplace, the community, and the environment and seeks to improve the overall quality of life in these communities. The movement, which emerged in the 1970s, criticizes government and industry for concentrating a disproportionate volume of toxics in workplaces and communities populated by poor people (Pellow and Matthews 2006, 136). It seeks to create sustainable communities—places with well-paid and safe jobs, good schools and recreation facilities, good housing and health care, and democratic decision making, all operating in communities that value cultural and biological diversity and distributive justice (Pellow and Park 2002, 4). Put simply, sustainability offers the promise of pollution prevention plus labor empowerment.

Scholars studying Silicon Valley (see Pellow and Park 2002; Rhee and Sadler 2007; Smith, Sonnenfeld, and Pellow 2006) identify a number of organizations that are pursuing labor empowerment in the area. These are not narrow, vanguardist organizations that focus first and foremost on a prolabor (or pro-trade-union) agenda. Rather, they are decentralized, innovative coalitions that are concerned about labor rights as human rights, and find common cause between workplace concerns, community concerns, and environmental concerns. They start by making the link between the environment and human health, arguing that unsafe and unhealthy workplaces result in unhealthy communities.

The Santa Clara Center for Occupational Safety and Health, known as SCCOSH, was the first organization to make clear the links between toxic workplaces and contaminated communities. Established in 1978, it combined two projects set up to monitor Silicon Valley industries. These groups started a support group for injured workers, helped to found an occupational clinic at a local medical center, created a hazards hotline for electronics workers, and launched a campaign to ban the chemical trichloroethylene (TCE) as a reproductive hazard. Pellow and Matthews quote one of the founders as stressing the workplace-community-environment links underpinning the coalition. She says that what made the fight to ban TCE noteworthy was the critical fact that "its origins and focus was always the workplace. If hazards faced by workers are not made a priority, we will all suffer the consequences" (2006, 136). Since the late 1990s, SCCOSH has been focusing more attention on the needs of the immigrant women who do the bulk of the work in assembling computers, components, and printed circuit boards. For example, one of its projects provides technical and practical information about hazards associated with these jobs, "and the skills needed to negotiate with management regarding these issues" (2006, 137).

One of the most prominent organizations to arise out of SCCOSH is the Silicon Valley Toxics Coalition (SVTC), which was created in 1982 and grew into an organization whose mission is to protect residents from toxic exposure in their communities. The group was triggered by concern about toxic chemicals leaking into the water table from underground storage tanks at IBM, Fairchild Semiconductor, and other companies. From its outset, the group made a conscious decision to work with the labor movement and SCCOSH to protect high-tech workers who suffer double exposure to toxic chemicals—in the workplace and in the community. Byster and Smith (2006) note that its board of directors traditionally includes representatives from the community, labor unions, and faith-based organizations—the types of organizations identified by scholars who study social movement unionism as typical partners in these movements, and as vital to their success. The toxics coalition has helped create national "right to know" laws that require companies to disclose information about the chemicals they discharge into the community. It has also set its sights well beyond the valley, launching a national network in 1990 known as the Campaign for Responsible Technology. In the late 1990s, the group helped create the International Campaign for Responsible Technology, whose goal is to "ensure that the high-tech industry and governments become accountable to its host communities and people and use the best practices to improve health and safety and reduce environmental impacts" (SVTC 2008). The international campaign serves as a clearinghouse and resource center for technical advice, organizing strategies, and information (Byster and Smith 2006, 118).

Some groups active in the area focus specifically on immigrants, including the Asian Immigrant Women's Advocates and the Labor Immigrant Organizing Network. The former works with women in the garment, hotel, restaurant, and electronics assembly industries, and emphasizes teaching them about labor and immigrants' rights. The latter serves as a vehicle for labor organizers to share recruitment strategies, pool resources, and lobby together for immigrants, including those in the United States illegally (Fan 2000). In addition, various "worker centers" throughout the valley pursue labor and community empowerment. Meanwhile, the South Bay Labor Council has what Gapasin and Wial (1998) would call a transformative orientation, fighting to give non-unionized workers access to benefits they can't receive from the high-tech firms that employ them. The council operates a think tank, Working Partnerships USA, which produces research reports and policy documents. The group has been very active at the local political level, helping to convince Santa Clara County to pass a strong "living wage" ordinance and to create a children's health initiative that provides health insurance to more than seventy thousand uninsured children in the area. Pellow and Matthews characterize the South Bay Labor Council as "closer to a 'worker center' model of organizing than a union" (2006, 137).

For the low-wage, immigrant information workers in Silicon Valley, participating in social movements that pursue common cause on issues like labor rights, health, and the environment is a feasible route to improved working conditions. These efforts emphasize the "social movement" dimension of social movement unionism. Together, these groups have led mobilizations to empower women and youth workers, to provide health-care access for temporary workers, and to pursue employer liability for a host of wage violations and for the impacts of chemical production and exposure. According to Pellow and Matthews, "these campaigns remain the best hope for strengthening labor rights and environmental justice in Santa Clara Valley" (2006, 138).

Information workers in other areas have also been experimenting with non-traditional ways of organizing or improving the lot of workers. Some, like Alliance@IBM, concentrate on a specific employer. Alliance@IBM is a local of the CWA that represents professionals at the computer giant, but not as the bargaining agent for them. Instead, it seeks to give workers a voice in shaping policies that affect pensions and health-care benefits. With a membership in the hundreds, it does not have enough support among the workers to gain bargaining-agent certification. But it is able to draw on the parent union's expertise to launch and run campaigns, legal challenges, and shareholder resolutions on issues like pensions, outsourcing, and workplace safety.

Others focus on contingent workers, including both creative and technical labor. The Freelancers Union, a national nonprofit organization, grew out of the group Working Today, which was founded in 1995 to provide benefits to people working in the New York City electronics district known as Silicon Alley. Membership is free of charge and open to all freelancers, consultants, independent contractors, temps, part-timers, and the self-employed, but members pay fees for the services they choose. Today, the group is able to offer group health insurance for members and their families in thirty states. By 2006, thirteen thousand of its members had purchased health care insurance and the union had grown to thirty-seven thousand members with annual revenues of $38 million and $4 million in funds for advocacy (*Economist* 2006). By the end of 2007 the union passed the forty-thousand-member mark and received national attention when freelancers walked off the job at the music video channel MTV after its parent, the media giant Viacom, approved a cut in benefits. Assisted by the Freelancers Union, picketing workers won a restoration of benefits and called attention to the plight of so-called permalancers who, like the permatemps at Microsoft, perform near full-time work for part-time wages and minimal benefits (Stelter 2007). The Graphic Artists Guild—representing people who work in illustration, graphic design, photography, cartooning, web design, multimedia, and other forms of design—combines elements of a professional association with trade unionism. It offers workshops that improve members' skills. But it also runs a legal defense fund and acts as an advocate for artists, particularly on the issues surrounding copyright. The National Writers Union, which participated in the early meetings that founded WashTech, gives members advice on freelance contracts and on asserting or protecting copyright. It also runs a job hotline and a campaign to get employers to hire a union writer.

In Canada, the Communications, Energy and Paperworkers union has organized a freelance writers' local, working in cooperation with a professional association, the Professional Writers Association of Canada. In December 2006, the fledging unit announced it would urge freelancers to "just say no" to a new freelance contract being handed out to writers at Sun Media, a subsidiary of the Quebecor newspaper chain. It has also tried to fight for higher freelance rates, protection of intellectual property, and better benefits for freelancers. By early 2007, it had roughly 350 members and had launched a recruiting drive in Toronto; Winnipeg; Ottawa; London, Ontario; Edmonton; Calgary; and Montreal. Plans for 2008 included a job board, described in the union's newsletter as a "hiring hall" for freelancers, and a survey of writers to find out who really pays what for freelance work (CFU 2007).

Finally, building on the freelance writers' movement and demonstrating that practically every form of new media, from the telegraph to the Internet, has given rise to labor agitation, in August 2007 an annual convention of bloggers hosted a panel on "A Union for Bloggers: It's Time to Organize."

The panel featured speakers who are members of a national coalition aiming to develop a labor organization for bloggers. Organizers hope a labor group will not only showcase the growing professionalism of web-based writers, but also the importance of their roles in candidates' campaigns. Blogging has become increasingly popular in the United States, where roughly 8 percent of Internet users have created their own blogs or online journals. A union or worker association might help bloggers receive health insurance, carry out collective bargaining and set professional standards. "It would raise the professionalism," said Leslie Robinson, a writer at Colorado Confidential.com. "Maybe we could get more jobs, bona fide jobs" (Heher 2007).

Promoting Information-Worker Rights

This chapter has described a range of approaches taken by unions, labor councils, labor-affiliated worker associations, and social movements to pursue, protect, and promote the labor rights of information workers. Some have been more successful than others, but all seek to give voice to workers, and to rebalance the increasingly inequitable power relationship between capital and labor.

The various campaigns arise as much out of necessity as out of choice. The business unionism model that has dominated labor relations in North America for decades is reaching the limit of its capabilities. Business unionism provided living wages for North American families, and it continues to do so for many workers. But it was built on a foundation of national companies and bricks and mortar factories. When the postwar social accord began to break down in the 1970s, the labor movement found itself having to adjust simultaneously to an increasingly hostile domestic environment and to increasing job loss resulting from factory closings, outsourcing of components, and offshoring of work. The unions also had to confront their own failures to organize beyond their traditional base—failings that became more apparent as trade union density, especially in the manufacturing sector, dropped precipitously. The rise of information or immaterial labor, bringing with it the possibility of flexibility in products, work processes, and markets to a degree never experienced before, has forced unions to rethink the ways they do business.

Some labor relations theorists have proposed value-added unionism as a way to revive the U.S. labor movement. This notion is based on the idea that unions should reinvent themselves as organizations which pursue goals that are of mutual interest to labor and capital. Rather than taking a confrontational stance, they believe labor should participate in partnerships with management. These would draw on labor's expertise in organizing workflow and allow employers the flexibility they claim to need. Attaining this goal requires unions to convince employers that working with a union is preferable to trying to fend off organizing campaigns or to decertify existing bargaining units. This may be feasible in certain conditions—especially in cases where a strong and imaginative union is already in place. But the idea raises questions about the fundamental relationship between labor and capital, and about the ability of value-added unionism to organize the unorganized. Other theorists have proposed social movement unionism as a path to renewal. This is based on the idea that labor needs to return to its roots as a social movement. Doing so requires finding and pursuing common cause between workers and the broader community. After an extensive review of the scholarly literature on both forms, Nissen concludes that of the two, social movement unionism "has a better chance to promote union renewal and upsurge" (2003, 148). Our research on information workers supports that conclusion.

It is clear, however, that none of the information-worker organizations discussed in this chapter fits the "ideal type" of social movement union described by Robinson (2002). This suggests that social movement unionism, especially in the evolving information sector, is itself very much a work in progress. The North American versions of social movement unions for information workers tend to be reformist rather than radical or revolutionary. The reforms they seek—such as legislation on outsourcing in the case of WashTech; or fuller disclosure of environmental risks in the case of the Silicon Valley social movements; or better contracts for freelance writers, graphic designers, or artists—are not limited to narrow workplace concerns. But nor are they aimed at overthrowing the social order. Nonetheless, each of the organizations analyzed here displays several characteristics of social movement unionism. For example, all exhibit a high degree of autonomy from the state, political parties, and employers. Membership in the information workers' associations is largely voluntary and open, and in some cases free of charge. This, combined with the evolving emphasis and orientation of the work these associations do, suggests strongly that these groups are accountable and responsive to their memberships. All the organizations work, to greater or lesser degrees, with organizations in the community, including religious or antipoverty groups, and groups that conduct research into social issues. Several of the organizations are affiliated with labor unions: the National Writers Union with the UAW; WashTech and Alliance@IBM with the CWA; the Canadian Freelance Union with the CEP. But for the most part these union-affiliated associations do not engage in traditional collective bargaining or business unionism. To a large extent, this is because they are not in a position to be certified as the collective bargaining agent.

Nonetheless, the larger unions believe it is worthwhile to continue investing in these organizations, and in others like them. In some cases, this reflects a tendency on the part of union leaders to refocus their vision of their own organizations. For example, Peter Murdoch, vice president of the CEP, would like to see his union operate for media workers the way Canadian teachers' and nurses' unions operate for their members: as recognized defenders of the profession and the values it embodies, as critics of government, and as advocates for a better society, and not just as bargaining agents (McKercher 2002). In other cases, the parent union sees these experiments in alternative forms of worker associations as worth pursuing for their own sake, especially in a changing economy. Says George Kohl, director of research for the CWA: "We're willing to underwrite any kind of experiment that comes forward in that sector that we think is a seed. You have to plant a lot of seeds before anything germinates" (Interview with authors 2007). Either way, these worker associations offer unions a chance to reach out to potential members and the communities in which they live, and they offer workers an opening into the resources and knowledge base of communications workers' unions. Asked whether WashTech would eventually be better off without the CWA, organizer Marcus Courtney replied, "Oh, God, no. There's just no way . . . [The CWA] helps provide a framework, a structure, ideas, information, advice, funding, trying to figure out how to do this work, how to organize this industry, how to build capacity, expertise, all that kind of stuff. There's just no way people could do it independently" (Interview with E. Brophy 2005).

Other worker associations representing information workers are not affiliated with a single trade union. Instead, they tend to operate with local labor councils or with community groups to build common cause. In some cases, such as the Creators Federation or the Freelancers Union, these tend to be specific to the needs of creative workers, such as protection of copyright or access to health insurance. In others, as in the case of the various associations in Silicon Valley, these tend to focus on environmental justice or social justice, cementing the link between labor rights and human rights.

Given the rise of precarious employment, especially in the information sector, we are likely to see more efforts to expand and consolidate the phenomenon of social movement unionism. Some, like WashTech or the Canadian Freelance Union, will draw on support from traditional trade unions to do nontraditional organizing, and in the process, perhaps, this will contribute to revitalizing the labor movement. We are also likely to see increasing efforts at building labor-community coalitions around the issues facing information workers, including living wage campaigns and campaigns to protect North American jobs. As we show in the next chapter, this is especially likely given the growing significance of the general upheaval taking place in the global division of labor and the specific and often contentious debate over the offshore outsourcing of information and communication work.

Chapter Seven

Workers in a Changing Global Division of Labor

Trade union convergence and social movement unionism are important developments in organized labor in the United States and Canada. But in a global economy, there are limits to what workers and their organizations can do at the national or even binational level. The development of communication and information technology has made it easier to create and manage transnational corporations. Government policies promoting trade liberalization provide strong incentives for national companies to establish markets abroad. Increasingly, these include markets for labor in both the industrial and knowledge sectors. China has attracted a great deal of attention as the world's factory, and anyone who has visited Hong Kong or the nearby manufacturing hub of Shenzhen on the mainland cannot help but literally breathe in the consequences. North American shoppers who buy deep-discounted toys, clothing, and practically every other material product at Wal-Mart or Target are hard pressed to find anything not made in China. Meanwhile, knowledge workers who labor on immaterial products are finding their work moving to India and other low-wage centers. And it is not just call centers employing tech-support workers that are sprouting up in what would once have been considered exotic locales. Jobs in journalism, health-care delivery, legal services, advertising, film production, and education are also affected.

In some respects it is more difficult to move knowledge or service labor abroad than manufacturing labor. That's because information labor requires language skills, education, and cultural knowledge not required of factory workers. On the other hand, immaterial products do not require container vessels to carry them across oceans. Advanced telecommunications systems do the job efficiently and inexpensively. With technical, language, and cultural training spreading around the globe, many of the barriers that protected knowledge jobs in North America are breaking down. It is therefore little wonder that the export of knowledge-worker jobs is on the rise.

This chapter takes up these developments with a specific focus on what workers and organized labor are doing about them. It also challenges popular views about outsourcing and suggests important underreported tendencies. In doing so, it aims to deepen the understanding of new developments in the global division of information and communication labor. The chapter includes two case studies that provide insight into different ways labor is approaching these issues. One looks at the response of American media unions to outsourcing jobs in the video and film industry. The other examines the rise of a global labor federation that aims to address the problem of outsourcing labor in the knowledge, media, and information industries. The chapter concludes by addressing the wider political, economic, and cultural significance of these developments.

Outsourcing is one tool in the business-led neo-liberal agenda that has transformed the social contract of the 1950s and 1960s. It refers to shifting work from in-house production to outside companies, generally in the name of productivity growth. The goal is to create leaner, more efficient companies with smaller core workforces, dealing on a contract basis with a range of goods and service suppliers. As we show, most of this occurs within the home country and within the developed world, though the proportion of offshore outsourcing to developing countries is rising dramatically. Regardless of where the work goes, however, it means a loss of job security for workers in the home country and increasing precarity in wages and benefits. Because outsourcing is part of a wider business trend that has also attacked the social policy instruments that protected working people, it has been all the more difficult for workers to mount a successful defense. In contrast to the social democratic vision of full employment and the optimization of all citizens' well-being, the neo-liberal state emphasizes ways to improve the business climate regardless of the consequences for employment and social welfare (Harvey 2006). Ironically, emerging liberalized economies such as India have recognized that massive intervention by the state is needed in infrastructure development, training, and education, to prepare the country's citizens for a global knowledge-based economy (Deshpande 2006). With 50 percent of its population off the electricity grid and 40 percent illiterate, India has a long way to go. It increasingly faces the challenge of squaring its neo-liberalism with the need for massive public intervention to provide basic services (Kapur 2007; Sengupta 2007).

Admittedly, the flight of capital and jobs is nothing new. New England's nineteenth-century textile mills, once models of corporate planning and business paternalism, now house museums, condominiums, and boutique malls.

The jobs are long gone, first to the southern states in the United States and now to Asia. But a case can be made that in an information society, something has changed. It is becoming easier and cheaper to outsource service work and creative labor, not just manufacturing jobs. Meanwhile, resistance is more difficult because business has rarely been stronger or more able to move.

It is also important to keep in mind that the lines separating manual from mental labor, or manufacturing from knowledge work, are often blurred, and that this has a bearing on how we understand outsourcing. Many companies try to sharpen that line by outsourcing what they believe to be the manual or manufacturing component, keeping the critical knowledge component—namely, the brand—at home. But that can be a lot more complicated than it looks. Take the case of Thomas the Tank Engine. Britain's HIT Entertainment owns the brand. The RC2 Corporation, which is based in Illinois, manufactures the Thomas & Friends toys in China. When a company moves production to China, it is able to distance itself from the labor process. That makes it more difficult to investigate working conditions and safety rules. Moreover, when labor problems are discovered, the multinational firm is able to claim ignorance of the practices and sharply distinguish itself from the Chinese manufacturer. In essence, companies get cheaper production without the bad publicity—at least, in theory.

In June 2007, RC2 announced a recall of 1.5 million toys after learning that the bright red and yellow paint decorating the toys contained lead. Lead-contaminated paint is toxic and chewing on the toys could seriously harm the children playing with them. Though RC2 announced the recall, the items recalled were not described as RC2 toys but as Thomas & Friends toys (see, for example, Jennings 2007). In other words, the brand sticks to the product in spite of all efforts to keep them separate in production. Things got worse a few months later, when RC2 announced a second round of Thomas & Friends recalls. Ironically, one of the items this time was a toy RC2 sent to customers after the first recall as a bonus "gift." As a *New York Times* columnist put it, "That's right. To make up for putting lead in its toys, RC2 sent children . . . a free lead-contaminated toy!" (Leonhardt 2007a). Leonhardt argued that the companies weren't simply unlucky: either out of carelessness or "a misplaced cost-cutting zeal," they did not take care of business. And it has cost them dearly. The companies have had to come up with ways not just to ensure safe manufacturing of the toys, but to rebuild customer confidence in the brand.

At around the same time, U.S. authorities recalled toothpaste bearing the Colgate brand because it contained a toxic chemical used in automobile antifreeze. The toothpaste was produced in South Africa and the Colgate brand was counterfeit. Nonetheless, Colgate-Palmolive paid heavily in bad publicity. In this case, the shadow world of factories outside the circle of rich nations simply claimed the brand, despite company efforts to wall it off.

Both the Thomas and Colgate recalls are examples of how the knowledge, design, idea, or brand element sticks to a manufacturing process in costly and dangerous ways. We can speak of the difference between outsourcing knowledge and manufacturing labor, but it is clear that these forms of production are mutually constituted. This means they can also be mutually contaminated.

Popular Views on Outsourcing

In a special report on outsourcing in 2006, *BusinessWeek* noted that work processes in virtually every department of a corporation can be outsourced or managed to some degree offshore. It produced a graphic of "the modular corporation" that represented some of the biggest sectors in terms of global spending in 2005. The numbers were estimates, and it was not clear how they were calculated. Nevertheless, they are impressive. And they offer some understanding of the relative values of different components of business that make up the offshore outsourcing industry. According to the *BusinessWeek* report (Engardio 2006, 55), the biggest sectors in 2005 included

- *Logistics and Procurement*: just-in-time shipping, parts purchasing, and after-sales repairs: $179 billion.
- *Manufacturing*: contract production ranging from electronics to medical devices: $170 billion.
- *Information Technology*: infrastructure, software development, website design, tech support: $90 billion.
- *Customer Care*: call centers for tech support, bill collection, bookings, and reservations: $41 billion.
- *Engineering*: testing and design of machinery and electronics: $27 billion.
- *Finance and Accounting*: billing, financial, and tax activity, accounts payable: $14 billion.
- *Human Resources*: training, payroll, and benefits administration: $13 billion.
- *Analytics*: market research, financial analysis, and risk analysis: $12 billion.

Given the massive amount of money fueling the outsourcing industry, labor fears massive loss of jobs. This fear is especially acute concerning jobs in the knowledge sector, the segment of the economy that holds the most promise for educated workers seeking good wages and creative work. Labor supporters propose legislation to significantly regulate outsourcing, including ending the growing practice of outsourcing government jobs and requiring call-center workers to identify their location. WashTech, the Communications Workers of America (CWA) local that has been tracking the loss of U.S. jobs since 2000, has led the charge to stop U.S. high-tech companies from sending jobs overseas. But not all trade unionists agree with calls for strict regulation. Andrew Stern, the head of the Service Employees International Union, believes that this approach is outmoded and futile. Instead, at a March 2005 conference of information-technology executives in Silicon Valley, Stern called on management and labor to unite in creating income-replacement and retraining programs for skilled workers (Langberg 2005). This position draws from a growing body of research that proposes new approaches to worker protection in the digital era, including continuity in wages, ongoing training, transferable skills, clear ownership of individual human capital, and portable health and retirement benefits, all of which would complement the traditional social safety net set up to help workers in the industrial era. These new approaches would also require a new form of trade unionism, one that emphasizes mutual aid and lobbying for legislation that is good for workers and deemphasizes firm-by-firm collective bargaining (Stone 2004).

The standard business argument in favor of outsourcing is the defense of free trade. According to this view, a staple of economics from Adam Smith on down, open markets for goods and services, capital and labor, permit the efficient allocation of resources. Nations and regions can concentrate on what gives them a comparative advantage. Naturally, jobs are lost here and there, but overall productivity increases and so does wealth. Rather than restrict outsourcing by businesses, according to this view, policymakers should encourage companies and their employees to work smarter by focusing their time and energy on creativity, innovation, and intelligent work. A popular version of this line of thinking is contained in the "Cirque du Soleil model," named after the Quebec-based company whose success has come not from producing a better circus but from reinventing the circus (Holstein 2005a). No longer defined by elephants and three rings, the Cirque is part theater, part ballet, part gymnastics, and part sound and light show. In addition, it is pitched at adults, not just children. According to this view, companies should be reinventing products from the ground up, exploding brand expectations, and redefining audiences and markets. It is hard to deny the Cirque du Soleil's success. In 2007 alone, close to 10 million people were expected to attend a Cirque performance and the company planned to perform fifteen different shows around the world, including permanent shows in Las Vegas. But the Cirque's success, which has now reached near-mythic status in the business press, has not created large numbers of jobs: the Cirque employs roughly thirty-eight hundred people (Cirque du Soleil 2007) and the traveling circus business as a whole continues its long decline. There is no necessary connection between business creativity and job creation.

In addition, just as some cracks have emerged in the trade unionist position on outsourcing, so too have disagreements emerged among the supporters of the practice. The most notable came when Nobel laureate Paul Samuelson questioned the benefits of liberalized trade and of outsourcing. The intensity of the issue was made abundantly clear when two senior economists submitted rebuttal articles to the same journal in which Samuelson's not-yet-published work was to appear (Bhagwati, Panagariya, and Srinivasan 2004; Lohr 2004).

Each of these popular views on outsourcing contains a basic truth, but each also oversimplifies a complex process that reveals significant changes in the international division of labor.

Outsourcing Knowledge Work Offshore

Most outsourcing takes place within the home country, including the fastest-growing form: business process outsourcing (BPO). This refers to transferring core or peripheral activities to an outside body, including call centers, accounting and finance, human resources and transaction processing (Centre for Education and Communication et al. 2006, 13). Offshore BPO accounted for only 1 to 2 percent of all BPO outsourcing in 2003–2004, but by 2007 that number was expected to grow to about 14 percent. That still leaves the vast proportion of this outsourced activity within the nation where the business is headquartered.

The head of the Information Technology Association of America, the trade association for the industry, estimated that about 4 percent of outsourcing activity was sent offshore in 2005 but the figure is expected to rise eventually to 40 percent (Koch 2005). Although the information is scattered across business reports and surveys of business plans, outsourcing of service work is clearly a growing phenomenon (United Nations Conference on Trade and Development [UNCTAD] 2004, Tables IV.3, IV.6). A frequently cited estimate is a Forrester Research report which concluded that between 350,000 and 400,000 service jobs had been lost by 2002 and anticipated a loss of up to 3.3 million by 2015 (McCarthy 2002). However, the report has been criticized for its methodological and empirical shortcomings (Roach 2005). In March 2005, the research firm Gartner reported that whereas only 5 percent of IT jobs in the United States were currently outsourced, by 2015, 30 percent of such jobs in the developed world could be outsourced (McDougall 2005). A report from the McKinsey Global Institute argued that 160 million service jobs—approximately 10 percent of all jobs worldwide—could be moved to remote locations because they do not require contact with customers, local knowledge, or complex involvement with the rest of the business (Porter 2005).

Some reports acknowledge the lack of good data but anticipate considerable growth anyway (Mann 2003). One reason why it is difficult to find good data is that companies are secretive about providing it. Governments, meanwhile, don't push them to report it. For example, the government of India does not gather its own data on employment in the information-technology-enabled services field. Rather, it relies on an industry lobbying group to provide it with information on overall employment. Moreover, in the United States, companies are not required to provide information to government agencies and, in fact, a growing number of firms are requiring their foreign partners to sign nondisclosure agreements, penalizing them if they or their employees divulge that they are working for a foreign company (Centre for Education and Communication et al. 2006, 20).

As outsourcing white-collar work moves up the value chain, discussion shifts from business process jobs to knowledge process jobs. Those shifted abroad have included credit, equity, and investment research work, as well as patent filing, intellectual property, and asset management. India alone is expected to need fifty thousand people in each of the next five years to meet the demand for outsourced higher-value knowledge process work, a growing trend in the subcontinent's economy (Reuters 2007; Teena 2006; *Telegraph.co.uk* 2007). In 2005, about three hundred companies were conducting knowledge process outsourcing in India, employing about sixty thousand people such as chartered accountants, science graduates and post-graduates, MBAs, market researchers, lawyers, engineers, and doctors. China, which produces four hundred thousand engineering graduates each year, many of them in computer studies, is an increasingly significant destination for such work. Microsoft began setting up computer laboratories in China in 1998 and has expanded the use of Chinese software engineers. In 2004 General Electric opened a large research center in Shanghai and was expected to employ twelve hundred researchers within a few years (Buckley 2004). So while it is hard to assess the future prospects for outsourcing with great precision, two trends are clear: expansion in numbers and movement up the job ladder. One expert offers this sober conclusion: "The bottom line, as I see it: We're largely flying blind in assessing the current and prospective magnitude of this important transformation in the U.S. labor market. My gut instinct tells me that this trend—like most IT-enabled developments in the past decade—is likely to proceed at a much faster pace than the consultants believe" (Roach 2005, 30).

Regardless of whether it is business processing or knowledge processing work, the bottom line is a key reason for outsourcing. The United States imports about $5 billion annually in software from India and cost is the major driving force (EPI 2004). About 70 percent of firms surveyed on why they send service work abroad identify savings in labor costs and from consolidating activity in a handful of specific foreign locations (UNCTAD 2004, 25). The United States is the primary originator of this activity, with two-thirds of all export-oriented information and telecommunications projects, including 60 percent of all call-center projects (2004, 26). Consider the case of IBM. In 2004 it employed 9,000 workers in India. By 2006 that number had grown to 43,000, including many who work on the design of specialized high-end chips in the company's first such production facility outside the United States (*Financial Express* 2007; Rai 2005b, 2006). By 2007 the company reached about 100,000 employees in India, some of whom were formerly employees of companies, such as AT&T, which outsource jobs to IBM, which then sends them to India (Courtney 2007b). IBM is, of course, not alone. In August 2005, WashTech reported that Hewlett-Packard planned to eliminate 14,500 jobs in the United States and has begun the process of having workers in Colorado Springs, Colorado, where it expected to eliminate 5,400 jobs, train their replacements to carry out service work in Mexico (Barr 2005; WashTech 2005). And in 2007, the major high-tech firm EDS announced an early retirement program for 12,000 of its American employees so that the firm could increase its total employees located in low-wage countries, mainly India, from 35,000 to 45,000 out of a total workforce of 136,000 (Associated Press 2007b). Another way for U.S. firms to position themselves for expansion in new markets is by buying local firms. In 2003 and 2004, many of the most important American high-tech companies purchased Indian companies. These included Hughes, which purchased Tenet Technology; General Electric, which bought the Engineering Analysis Centre of Excellence; Perot Systems, which took over Vision Healthsource; and IBM, which purchased Dasksh eServices (UNCTAD 2004, 160).

Outsourcing may also be carried out to strengthen management's position in labor disputes. For example, in 2005 the Telus Corporation, Canada's second largest telecommunications provider, locked out 11,500 employees across western Canada represented by the Telecommunications Workers Union. In preparation for the lockout, the company set up call centers in India and the Philippines to handle operator and customer service work. Moreover, when the company expanded its operations into central Canada, it hired an Indian subsidiary to carry out a major part of the technical work to integrate its Canadian services.

American businesses have taken advantage of U.S. immigration laws, which permit bringing in skilled technical workers on a temporary basis when the federal government determines that there are insufficient numbers of U.S. workers. WashTech and its allies maintain that companies use this system of H-1B visas to lower the wage bill and train foreign workers, who return home only to continue working for the company in its outsourced operations. One business executive admitted: "They're using it as a purely temporary means to bring in rank-and-file employees who will work for less money. They're not even bothering looking for American workers. This is pretty rampant in the IT sector in particular. These workers then gain experience in the U.S. on the latest technologies, interfacing with customers, and then they are able to take that back to their home country" (Holstein 2005b).

There is growing evidence of outsourcing media work, though much of it is anecdotal. The concern about outsourcing work in this sector of the knowledge industry has grown with layoffs in the U.S. media sector, where in 2006 nearly eighteen thousand media employees lost their jobs—an increase of 88 percent over the previous year. In December 2007, media employment in the United States hit a fifteen-year low (Johnson 2008). Some of the companies doing the downsizing are among the industry's largest, such as Time Warner's AOL, which began cutting five thousand employees in December 2006, or 26 percent of its workforce (Becker 2007). Reuters has taken the lead in the news business by moving jobs to India. In 2005 it announced that fully 10 percent of its workforce, some twelve to fifteen hundred jobs, were on the way to Bangalore. Most of these jobs involve sorting stories for distribution through the company's numerous subscription services (Timmons 2004). About fifty of the jobs are in editing and writing. Reuters also plans to move its photo-editing desks from North America to Singapore (Chepesiuk 2005). A 2007 story grabbed attention by proposing that outsourcing may, in some cases, offer cost savings even in coverage of local news. An online news site in Pasadena, California, used an online ad to recruit reporters in Bangalore and Mumbai to cover local Pasadena city council meetings. The reporters would watch the council meetings on the Internet, writing stories about what happened at the meetings and contributing depth pieces every week. The editor agreed to pay the journalist in Mumbai US$12,000 and the one in Bangalore US$7,200. The announcement attracted a great deal of attention from other journalists, much of it negative. A few days after announcing he had hired reporters, the editor of the site said he was postponing the idea (Associated Press 2007a). In early 2008, Canada's Metroland West Media Group, part of a community newspaper and magazine group owned by the Torstar Corp., signed a contract to make Pressmart Media Ltd. of Hyderabad, India, and New York the service provider for its digital editions. Pressmart operates online publishing services in twenty-three countries (*Editor and Publisher* 2008).

Hollywood has outsourced many of the jobs that once fed the southern California economy, with animation work going to Asia (Day 2007) and film production to Canada (Elmer and Gasher 2005a). By 2007 film production had expanded globally, with eastern Europe taking an increasing share of the work. Some accounts document the beginnings of outsourcing in education, particularly the private tutoring business. Companies like Growing Stars and Career Launcher in India charge American students $20 per hour for tutoring, far less than the $50 or more that companies based in the United States can demand. In fact, these companies demonstrate their flexibility in the "homework outsourcing" business by offering American parents the choice of an Indian tutor or, for a higher fee, an American tutor. The transaction and the tutoring all take place over the Internet (Rai 2005a; see also Dhillon 2006).

The advertising industry has big plans for using outsourced labor in China and India to help take the industry to the next stage of development. Those who track the industry envision a transformation from thirty- or sixty-second spot ads to personalized advertisements, tailored to individual consumers. This development received a boost in 2007 when the global advertising conglomerate Publicis—which owns such leaders in the field as Saatchi and Saatchi, Leo Burnett, and Starcom and handles the accounts of Procter and Gamble, General Motors, American Express, and Hewlett Packard—purchased the pioneering firm Digitas for $1.3 billion. Publicis hopes to create a global ad network that uses offshore labor, primarily from India and China, to create thousands of versions of advertisements in digital form. Drawing on data about consumers, including their purchasing patterns, the network will use computer algorithms to decide which advertising message to show at which moment to every person who turns on a computer, a cell phone, or—eventually—a television set. Advanced design teams based in the West will develop ad templates that can be modified in many different ways by offshore workers who will feed them into the global network for dissemination to the appropriate media device. The company already owns a major advertising firm in China, whose annual growth rate of 20 percent a year dwarfs the industry average of 5 percent. The additional workers from Digitas will supplement an offshore labor force that already includes employees in Costa Rica and Ukraine (Story 2007).

Some suggest that the next major stage in the outsourcing movement will involve person-to-person services for consumers and small businesses. So when Adrianne Yamaki, an overworked management consultant in New York, tries to find more time for herself, she regularly outsources the routine aspects of her life—such as making travel bookings, hair appointments, and restaurant reservations and purchasing theater tickets—to a personal assistant service, in India. When Kenneth Tham, a high school sophomore in Arcadia, California, looks to improve his grades and his scores on standardized tests, he is tutored remotely by an instructor based in India. This individual speaks to him on a voice-over-Internet headset as Kenneth sits at his personal computer, going over lessons presented on the screen. According to Lohr (2007b), this globalization of personal consumer services is expected to become a $2 billion business by 2015 and Indian entrepreneurs, from small to large businesses, are beginning to invest in it. There are certainly major challenges, including cultural hurdles and the difficulty of establishing the economies of scale to overcome communication costs, but there have been enough success stories to make this worth closer attention.

As this brief discussion has shown, the pace of offshore outsourcing is increasing, and so is the variety of communication jobs liable to be sent abroad. It is certainly important for anyone who studies worker organizations to pay attention to the phenomenon and the job loss that accompanies it. However, it is easy to get bogged down in competing numbers and anecdotal evidence, and to lose sight of the wider consequence of outsourcing. Huws reminds us that workers in the developed world tend to believe that the point of outsourcing is to eliminate their jobs. But this misses the crucial idea that a deeper purpose of outsourcing is to serve as a "disciplinary force" on workers:

> If workers know that the skills they have are also held by hundreds of thousands of other people around the world, then it is very difficult to organize around their unique occupational identities. And if they are aware that it would be perfectly feasible, technologically speaking, to move their jobs offshore, then this creates a potent disincentive to ask for improvements in wages and working conditions, or to refuse to take on extra tasks. Just the possibility that the job *might* move is enough to destroy their security and worker's bargaining power. (Huws 2006, n.p.)

There are exceptions to this rule, with some employers opting for retraining in lieu of outsourcing (Kramer 2007). But these exceptions are rare, and they mainly underscore the point Huws makes: the mere threat of outsourcing, which hangs over practically every industry today, is a powerful deterrent to reviving the labor movement.

Outsourcing within the Developed World

Outsourcing began as a process of sending work to other parts of an enterprise's home country, and that practice continues, even in the knowledge sector. So while Bermuda and the Cayman Islands advertise themselves as havens for the U.S. insurance industry that is eager to make use of skilled, cheap labor and take advantage of minimal taxation, the state of Vermont is up there with the offshore islands as a major destination for companies that want to reduce costs. They do so partly because Vermont has skilled labor and is close to major East Coast centers like New York and Boston. But in addition, the state has drafted a tax code that has attracted 560 U.S. firms—including Wal-Mart, Starbucks, McGraw-Hill, and, most recently, Wells Fargo—which have set up entities in the state to insure their biggest risks and liabilities, and gain a tax benefit in the process (Browning 2007). This advantage may prove short lived, but it is important to keep in mind as the attention turns to India and China.

It is also important to keep in mind that, to the surprise of some, the recipients of most outsourcing of knowledge work are developed countries in Europe and North America. In 2002–2003 over half the foreign call-center projects went to developed countries, with Ireland, Canada, and the United Kingdom in the lead. In 2001, the total market for all offshore services was $32 billion and fully a quarter of that total was accounted for by Ireland alone (UNCTAD 2004). Developed countries offer some cost advantages for U.S. firms, but also a supply of well-educated, skilled labor that is particularly important as companies use technology to extend the range of exportable services up the knowledge chain. Call centers require good communication skills and some knowledge of the culture of an exporting country's market. This helps to explain why English-speaking Canada, which provides some, but not a significant, cost advantage, is an important source of labor for American call-center firms. French-speaking and bilingual regions of Canada provide jobs for French companies as well. But education and skill are even more important for more demanding occupations that are increasingly exportable, such as software engineering, architectural design, financial analysis, diagnostic radiology, and legal services. While some of these jobs are going to India, many more are heading to Ireland, Canada, Israel, the UK, and various countries in Europe. India and China are often singled out as the bad boys of outsourcing, but Canada, Ireland, Israel, and other places have taken the bulk of the knowledge-industry jobs. At least, so far.

Indeed, we are observing a global dynamic that is more complex than the standard view that jobs get shifted from high-wage regions to low-wage regions. Jobs move everywhere, and precisely where they go depends on a variety of considerations. Some of these considerations—such as language, skill, and education—dictate a move within the developed world. In addition, some jobs move more than once—for example, from the United States to Canada and then to India. As an executive for a Boston-based software outsourcing company puts it, "In some cases we use Canada as a front end to India. We find that this takes away the issues people have with India" (Austen 2004).

India Takes a Lead

The emphasis on India and China as primary destinations for outsourced knowledge-sector jobs may be overplayed today, but there are good reasons to expect these countries to play a very significant role soon. Estimates, admittedly contested, suggest that between 3 million and 4 million service jobs may shift from the United States to developing countries by 2015, with 2 million in the financial services industry alone (McCarthy 2002). Consider the outsourcing of legal work, one of the major areas of knowledge processing labor. We are beginning to observe the shift of legal services work to India, with about thirteen hundred Indian workers providing services for U.S. lawyers generating about $52 million in annual revenue. Although there is concern about confidentiality issues, the revenue from this kind of work is expected to reach about $1 billion in a decade. The numbers today are small compared to the approximately 1 million U.S. workers employed as lawyers or paralegals. But because Indian law firms charge one-third of what their American counterparts charge, the cost savings offer the potential for significant expansion.

One of the more interesting and underreported developments is that historically poor countries are not just sources of outsourced low-wage labor, but are beginning to take a leadership role in the outsourcing process. A typical example is the India-based transnational company ICICI OneSource Ltd. It is part of the ICICI Group, a major global finance company based in Bombay. ICICI OneSource provides both local and outsourced services—including customer service, complaints resolution, and telemarketing—from call centers in India. But it also does market research and analysis from its Chicago office. In 2005 ICICI acquired a U.S. firm outside Buffalo, New York, that handles so-called late-stage credit card collections, or bills that are long in arrears, for U.S. credit card firms. Many Indian firms have established a presence in the United States by opening offices or branches there in the last few years. ICICI's decision to buy a U.S. business moves things to another level. Initially retaining the five hundred American employees and naming the company's president as its head of global collections, ICICI OneSource began the process of integrating the American operations into its international business. The significance of an Indian multinational using the president of an American company it has just acquired to serve as its global bill collector should not be lost on those who see outsourcing as little more than an extension of American neo-colonialism. The company is making the move partly because it wants to strike deals with U.S. and Canadian firms to provide outsourced services from its Indian locations, but also from the United States and Canada. According to the company's vice president of sales for Canada, that country's French-language capability, talent, and culture give it an advantage to service the Canadian bilingual market, as well as the United States and parts of Europe (Galt 2005). The border location of ICICI's first outright U.S. acquisition gives the company easy access to the Canadian market. Establishing a local presence is a particularly important consideration for outsourced work like late-stage bill collection, which benefits from an understanding of local conditions.

India's Infosys Technologies, which earned over $3 billion in 2007, is another rapidly growing firm in the global outsourcing business. Although the vast majority of the company's seventy-five thousand employees are Indians living in India, Infosys is developing a global system of back offices that includes Mexico, the Czech Republic, Poland, Thailand, the Philippines, and even low-wage regions of the United States. The company's outsourcing operations confound popular conceptions. One example involves an American bank that wanted a computer system to manage a loan program for Hispanic customers. The system not only had to work in Spanish but had to manage practices common among Hispanic clients, such as remitting money to families living abroad. The American bank thought an experienced Mexican team was necessary, but rather than pursue a Mexican vendor, or even an American one with Mexican connections, it hired engineers from Infosys, which had recently opened an office in that country. So in this case, outsourcing means that an American company pays an Indian firm to provide Mexican engineers (Giridharadas 2007b).

Tata Consultancy Services, Infosys's Indian rival, has also stepped up its outsourcing activity. By 2006, Tata had spread to Latin America, with operations in Brazil, Chile, Mexico, Argentina, and Uruguay. In Montevideo alone, it employed 550 programmers, trained and directed by Indians, who write code and run the computer systems of companies all across the continent (Friedman 2006). In May 2007 it announced a new back office in Guadalajara, Mexico, adding to the five thousand workers it already employed in Latin America. Wipro Technologies, another major Indian technology services company, has outsourcing offices in Canada, Portugal, Romania, and Saudi Arabia. In 2007 it opened a software development center in Atlanta to house five hundred programmers and is considering hubs in Virginia and Idaho to take advantage of what its chairman calls American "states which are less developed" (Giridharadas 2007b).

It is significant that revenues earned by Indian information technology services are double those earned in China, where the business is growing but the firms are just too small to win deals with the top international firms now (Frauenheim 2005). Large Indian companies like Infosys and Wipro are, however, planning to acquire Chinese companies to expand operations and establish a base in a key economy that will enable them to play an even bigger role in global decisions about outsourcing. Further deals like these are increasingly likely, and more of a challenge for China, now that the World Trade Organization has forced the country to phase out rules requiring foreign firms to undertake joint ventures with Chinese firms as a condition of doing business in China.

In another important move, Tata Consultancy Services has established a base in Vancouver, the better to compete with Canadian firms for U.S. outsourcing business and to train tech workers for its international customers.

Vancouver follows Toronto, Montreal, and Ottawa as the fourth Canadian city for Tata, which was the first Indian firm to surpass $1 billion in annual revenues. Vancouver is perhaps also the most important city for Tata because of its proximity to the Seattle information technology cluster, and especially to Microsoft's headquarters in nearby Redmond, Washington. In the company's own business lingo, Vancouver provides Tata with the "near-shore optics" that will enable it to more easily capture business in the United States. In a further reversal of the typical scenario, Tata has signed an internship agreement with Simon Fraser University, also located in the Vancouver area. The university's computer science students do internships with Tata and develop software for the company. With more than eight thousand employees in North America and plans to aggressively pursue business globally, Tata belies the image of India as the low-wage dumping ground for U.S. firms (Fong 2004). Admittedly, there are numerous reports of long hours, low wages, poor working conditions, and oppressive management of high-tech work in India (Kapur 2007; Kumar and Verghese 2004; Nandgaonkar 2005; Sinha 2004; Tollefson 2005; Upadhya and Vasavi 2008). But India's leading companies have also learned how to operate as global leaders in the knowledge labor arena.

One of the best examples of how far India has come in taking a lead in the international division of labor can been found in a feature of the H-1B visa story that has not received much attention. Most of the debate about the visa program has understandably swirled around Microsoft because the company and its leader, Bill Gates, have lobbied fiercely and publicly for expanding the program and eliminating the cap on the number of visas issued each year. But the list of companies requesting such visas in 2006 was led by Indian firms: Infosys was first, Wipro was second, and Tata was eighth. In fact, eight of the top twenty firms were based in India (Hira 2007). These firms want to bring "foreign" workers into their U.S. operations for a short period of time to enhance their global outsourcing operations. Specifically, the companies import visa workers to the United States to learn the jobs of American workers, so that they can apply them to offshored operations—a practice euphemistically known as "facilitating knowledge transfer operations." Moreover, visa workers are used to help coordinate offshore operations. Wipro has four thousand workers in the United States, twenty-five hundred of them on visas. About one thousand come into the U.S. each year and another one-thousand go back to India with enhanced skills to support the company's outsourcing operations. Meanwhile, the firm continues to purchase companies that do business in the United States. For example, in August 2007, Wipro picked up Infocrossing, a New Jersey company with nine hundred employees at five data centers providing IT services to medium-sized U.S. companies. In December that year, Infocrossing won a $400 million contract to provide services to Missouri's MO HealthNet (*Indiatimes* 2007).

In essence, Indian firms are using a U.S. base and a U.S. visa program, designed to support American companies, to enhance their global operations and accelerate the process of outsourcing high-tech jobs from the United States. No doubt, U.S. high-tech firms like IBM and Microsoft, and major U.S.-based outsourcing companies like Deloitte Touche, have benefited enormously from the H1-B program. In fact, Microsoft was so upset when a bill to expand the program in 2007 was killed that it set up a software center in Vancouver to take advantage of Canada's immigration system (Bishop 2007). The visa program has also helped IBM to manage its Indian operations, which employed fifty-three thousand workers in 2007, up from three thousand in 2002. (IBM's employment in the United States had dropped to 127,000 at the end of 2006.) But the presence of so many Indian firms on the list of companies seeking visas to bring people into the United States demonstrates the growing power of Indian firms in the international division of labor (Hira 2007). In a related development, Indian firms have also established a strong lobbying presence in Washington, working with some of the top U.S. consulting firms to put an American face and an American spin on their business activities, especially in the world of outsourcing knowledge labor (Giridharadas 2007a).

The strength of Indian companies has forced leading firms like IBM to restructure their own management and operations. As of mid-2007, leading Indian outsourcing companies enjoyed profit margins that surpassed 20 percent, twice those of IBM, which leads the American pack of technology services companies. IBM remains considerably larger than Wipro, Infosys, and Tata, but when Infosys, for example, reports quarterly revenue gains of more than 40 percent, as it did in the second quarter of 2007, big firms have to take notice. Indeed, the cochairman of Infosys likens U.S. technology services companies to Detroit automakers, who try to copy but inevitably fail to match the foreign competition (Lohr 2007a).

Some have suggested that the great irony in contemporary outsourcing is that Asia is finally taking economic revenge more than three hundred years after its development was stymied by the West (Monbiot 2003). Britain's industrialism succeeded after it destroyed the manufacturing capacity of India. For two centuries of colonial subjugation, India was forced to provide raw materials for Britain's industry but was prohibited from producing finished products that might compete with those of the UK. Now the flow of jobs has reversed, and India's knowledge sector is growing at the expense of the knowledge sector in Britain and other countries in the developed world. But as Mike Davis points out in his powerful book *Late Victorian Holocausts* (2001), this is more tragic than ironic, and "revenge" may very well be an appropriate word. He shows that in the last half of the nineteenth century, during the first major wave of globalization, a combination of British capitalism and climate change killed between 30 million and 60 million people in India and China in famines that were profoundly deepened by a free market ideology that refused traditional means of helping the needy, the sick, and the dying.

The End of the End of Geography

ICICI's move into the American heartland, and Tata's into the Vancouver market, provide evidence of another key and little-noticed dimension of the outsourcing issue. In spite of "end of geography" promises, which have promoted what amounts to a minor industry in mythmaking, place still matters and culture still counts (Mosco 2004). ICICI's decision to acquire a Buffalo firm and use it to expand bill collection operations in North America acknowledges the importance of location and culture. According to the company's head of North American operations, "For instance, late-stage credit card collection, basically those accounts that are 120 days to 180 days overdue, should be done from a U.S. location, not from an offshore location" (Galt 2005). This is principally because employees located near their market would be more aware of local conditions, such as plant closings or generally high unemployment rates, that might influence the success of the collection process. Similarly, Tata's choice of a Pacific Northwest base acknowledges that, even though information moves at the speed of light, successful companies have to physically set up shop in or near the center of their business activity. Just as Microsoft has to be physically located in Bangalore to take advantage of an important cluster of high-tech activity in the global economy, Tata needs to be in Vancouver. Additionally, Tata's action demonstrates the importance of culture in the knowledge industry. One of the primary reasons why Ireland, Canada, and Israel have been key participants in the outsourcing industry is that their language and culture are close enough to the major knowledge industry firms to make them reliable locations for the work, even when they are not the lowest-cost alternative. Tata locates in Vancouver for what it calls the optics, by which it means not just the appearance of being a Western firm, but also the opportunity to absorb, and be absorbed by, the culture of one of the high-tech industry's major regions. For many of the same reasons, companies are now exploring South Africa as a destination for outsourced work. Even though that country is more expensive than some of the prime low-wage targets, for years it has provided the base for many multinational firms and so it has a "home" feel. The universality of English and the additional political credit received for creating jobs in Africa are also factors (*Economist* 2005a). Moreover, companies are taking advantage of national connections in developing patterns of global location. For example, when a company started by Hispanic immigrants to the United States to provide businesses with computer tech support set up its coordinating office, it chose a Bogota location because the city could offer low-cost, skilled labor. A location in the same time zone as its primary service area, New York City, also mattered, as did the low turnover of its Bogota-based workforce (Hanc 2007).

Companies are increasingly turning to India and China to outsource work, including business and knowledge processing jobs, and low labor costs make a difference. But that is only a very simplistic explanation of a very complex process. Some companies find it advantageous to take outsourcing in another direction. When General Electric, which owns NBC, bought the Spanish-language network Telemundo, it began outsourcing television production from Mexico, where Telemundo was based, to Miami, which has become a center for Spanish-language productions once done primarily in Latin America.

Moreover, the decision to pursue outsourcing is no guarantee of its success. According to a 2005 report of the Conference Board, "fully half of all offshoring operations are destined to fail" (Koch 2005). The report cautions firms against rushing in without careful planning. A survey by Bain and Company concluded that firms are "outsourcing more and enjoying it less." In particular, it found that although 82 percent of large firms in Europe, Asia, and North America are making some use of outsourcing firms, and 51 percent are outsourcing offshore, almost half say that their outsourcing does not meet their expectations (*Economist* 2005b). Companies are learning, sometimes the hard way, that the nation-state and cultural obstacles remain, posing major challenges to the widespread belief that technology is creating a flat world (Friedman 2005; Moore and Hertz 2008). Meanwhile, insecurity about outsourcing is growing within the high-tech industry, even at higher levels of authority. A 2007 survey in the UK found that over two-fifths of software and web developers agree or strongly agree that offshoring is a threat to their jobs. This compares to less than a third of chief information officers and just over a third of board-level executives and IT managers. However, 23 percent of all respondents were uncertain about the impact on jobs, an increase of 6 percent over a 2004 survey. Such insecurity and ambivalence among technical professionals and managers suggests that the road ahead for high-tech outsourcing may not be as smooth as its strongest supporters anticipate (Lomas 2007).

Regardless of whether corporations succeed at outsourcing knowledge-worker jobs, the organizations that represent workers across the knowledge, communication, and cultural industries are looking for ways to respond to the challenge of outsourcing and to the seemingly relentless upheaval in the international division of labor. We now take up case studies that examine two of these responses to developments in the new global economy. The first addresses a key component of the cultural industry, where trade unions in the film and television industry have developed strategies to combat the problem known as runaway production. The successes and failures of this struggle contain important lessons for building labor power worldwide. The second considers the potential for a global labor response in the vast electronics information and services arena. Specifically, it explores the case of the Union Network International (UNI), a leading global union federation that has set out to bridge the many gaps that divide knowledge workers of the world. Its goal is to develop common strategies of resistance—in essence, to use labor convergence and social movement unionism to help knowledge workers of the world to unite.

Reining in the Runaways

We begin by looking at how the unions that represent mainly creative labor in Hollywood have fought against their own version of offshore outsourcing: the phenomenon known as runaway production. Although the term has several nuances, at its simplest level it refers to the idea of filming American stories, intended for an American audience, in another country. The Screen Actors Guild (SAG), a union of 120,000 actors, has been a key player in the struggle, fighting runaway production on a number of fronts. The guild has drawn on a range of discursive strategies, casting the problem as an economic, cultural, and patriotic issue, as well as a matter of traditional workplace jurisdiction. It has also employed a number of tactics, such as commissioning studies that allowed the unions to define the issue, lobbying for state and federal legislation, launching publicity campaigns, and attempting to extend SAG jurisdiction beyond U.S. borders.

It is important to address this sector because the production and distribution of culture is an increasingly valuable commodity in the global economy and a major means of forging individual and collective identities. Whether in Hollywood or Bollywood, people look to the cultural sector as a source of work. In addition, the film and television business is one of the major industries feeding the dream of striking it rich (Wasko 2004). This case demonstrates that many of the efforts undertaken by SAG have paid off. But the successes depend to an enormous extent on SAG's ability to build or to work in coalitions—with other unions, of course, but with employers and civic groups as well. Achieving solidarity is harder than it looks, and there have been missed opportunities along the way. The most significant is the guild's failure to merge with the American Federation of Television and Radio Artists (AFTRA). As we showed in chapter 5, a merger between the two unions would have increased labor power for performers working in film, television, and commercials. The repeated failure to achieve that merger has had implications in several areas—including, as we will show here, in the fight against runaway production.

Although runaway production is often seen as a phenomenon of the late twentieth century, the practice of filming "Hollywood" movies in other locales has been a matter of concern to the Los Angeles–based film unions since shortly after the end of the Second World War. The Screen Actors Guild, founded in 1933 as a centralized organization representing film actors working for the powerful and paternalistic Hollywood film studios, found itself swept up in a tide of changes in the entertainment industry in the early postwar years. A 1948 antitrust decision, known as the Paramount decree, forced studios to give up their control over theater distribution, and this led to a major restructuring in Hollywood. The number of independent producers rose from 40 in 1945 to 165 by 1957 and 1,500 by 1966. The proportion of films made by the seven major studios that once dominated Hollywood dropped from 80 percent in 1949 to 35 percent in 1958 (Prindle 1988, 69–72). The arrival of television, meanwhile, threatened the cinema business and launched SAG on a jurisdictional fight with the American Federation of Radio Artists (AFRA), a fight resolved in 1950 by giving the federation jurisdiction over live programming and the guild jurisdiction over filmed work. AFRA renamed itself AFTRA in 1952 in recognition of its expanded jurisdiction.

In an attempt to follow the work being filmed outside Hollywood, SAG began opening branches across the United States. Founding member Kenneth Thomson told the 1953 guild annual convention, "Although it did not set out to do so of its own volition, the Screen Actors Guild has now of necessity become a nationwide organization, with branches and membership not only in Hollywood and New York—but also in Chicago, Detroit, San Francisco, Cleveland, Cincinnati, Pittsburgh and Boston" (SAG 2007c). Indeed, by the 1980s, less than half of SAG's membership resided in its founding Hollywood branch (Prindle 1988, 124). But while the union could expand within the United States, it became increasingly frustrated watching producers leave the country in search of cheaper locations to film their work.

According to Prindle, by the end of the 1960s about half of the so-called American films were being shot in other countries, using American stars but local actors in supporting parts. "Repeatedly during the decade, Guild leaders harangued the industry about its exporting of jobs, lobbied Congress to pass tax legislation offering incentives to filmmakers to stay in America and made contract concessions to encourage domestic production. These efforts had no visible effect" (1988, 92).

By the 1980s and 1990s, the anger over foreign production had become more focused. The problem, as SAG and other unions saw it, was not runaway production in general, but runaway production in Canada. Canada had been developing its own film and television industry during these years with the aid of government support, ranging from tax incentives to subsidies. Filming in Canada offered a lot of attractions to production companies. Canada was foreign, but it was not exotic. Its landscape, cities, and citizens looked American. Toronto or Montreal could stand in for Chicago or New York; Manitoba for anywhere in the Midwest; and British Columbia could pass for California, Montana, or Alaska. Add to this an inexpensive Canadian dollar and the availability of tax incentives, and it was inevitable that the number of productions filmed in Canada would begin to rise.

SAG and the Directors Guild of America took command of the issue in 1999, commissioning a major study of runaway production from the influential Monitor Company, a global management consulting firm. The study sought to quantify the extent of the problem and assess its economic impact. The fact that the study was done by a third party—neither by the unions nor by the film producers—gave it a credibility it might not have had otherwise. The research could not be dismissed as union propaganda or a compilation of anecdotal material; rather, it was a serious examination of an issue by people with independent expertise. The unions hoped that this study would define the issue: it would clarify exactly what constituted a runaway production, and it would show who was benefiting from the phenomenon, and who was being hurt. They got their wish, and more.

The Monitor study defined runaway productions as films or programs that are filmed in another country but intended for initial release or broadcast in the United States. It determined that there were two types of runaway productions: creative productions, which depart for creative considerations or because the story takes place in a setting that cannot be duplicated in Hollywood, and economic runaways, which depart to achieve lower production costs (Monitor Company 1999, 2). It was the second group that posed the problem. And the problem was huge—big enough to affect not just Hollywood but the overall U.S. economy. The study estimated that economic runaway film and television production cost the U.S. economy a total of $10.3 billion in 1998, up from $2 billion in 1990. This included the loss of direct production spending, lost tax revenues, and multiplier effects that occur with significant job loss. The study found that the vast majority of those runaways, 81 percent, went to Canada, with Australia and the United Kingdom running a distant second and third. The labor impact of economic runaways was "profound": the loss of more than twenty thousand full-time-equivalent jobs in 1998, eleven thousand of them jobs usually filled by SAG members, six hundred jobs normally filled by Directors Guild members, and the rest by so-called below the line craft and technical workers (1999, 2–4). The study identified several causes of the problem. It noted that Canadian and Australian film crews and infrastructure had improved in recent years, to the point where there was no significant difference between the quality of Hollywood-filmed productions and those filmed in Canada or Australia. At the same time, the value of the Canadian, Australian, and UK currencies had declined. Tax credits were readily available, especially in Canada. Filming in Canada, it estimated, could cut production costs by 25 percent or more.

The report provoked significant reaction. Magder and Burston note that Canadians from both industry and government took issue with the financial analysis (2001, 216). The Directors Guild of Canada, for example, claimed that the Monitor report exaggerated the amount of money coming to Canada by a factor of three or four. Magder and Burston describe the report as a "dishearteningly simplistic and one-dimensional intervention in a much needed debate on the globalization and digitization of film and televisual production" (2001, 226). They argue that the "blame Canada" frame had little analytical purchase. It was, however, politically effective for the two unions that commissioned the report. The report gave SAG and the Directors Guild ownership of the issue. They had become what Hall and colleagues (1978) would call the "primary definers" of the economic problem, and also of the stories and debate the report would generate. Everyone else—the federal and provincial or state governments in Canada and the United States, producers in both countries, labor unions in Canada, and so on—was put in the position of commenting on or reacting to the study. Again to use terms from Hall and colleagues, these groups became the secondary definers. They might take issue with the study's numbers or how they had been tallied, but the notion that runaway production was an economic problem quickly became the dominant interpretation, and it set the terms of reference for further discussion of the matter. According to the *Hollywood Reporter*, Canadian producers conceded privately that the report had become the "bible" on the issue (Vlessing 2004).

With runaway production defined as an economic problem, the Screen Actors Guild and the Directors Guild began to concentrate on a solution. The way to counter foreign incentives that drew production out of the United States, they concluded, was for the United States to offer incentives to keep it at home. But the guilds knew it would take more than a couple of unions to successfully make the case for incentives. It would require the assemblage of a broad coalition representing every aspect of Hollywood. The result was the Runaway Production Alliance, composed of nineteen organizations with a stake in the issue. Spearheaded by the Directors Guild, it included labor unions (SAG, AFTRA, the Teamsters, the International Alliance of Theatrical Stage Employees or IATSE, the Writers Guild, and so on); professional or trade organizations (the Producers Guild of America, the Association of Independent Commercial Producers, the Hollywood Post Alliance, the Association of Independent Video and Filmmakers, and the Caucus for Television

Producers, Writers and Directors); marketers, distributors, and promoters (the Independent Film and Television Alliance, the Motion Picture Association of America); organizations representing talent agents and musicians; and the Academy of Television Arts and Sciences, the group that runs the Emmy Awards (AFTRA 2004).

SAG views forming alliances across the industry to lobby Congress on specific issues as a good idea. Pamm Fair, the deputy national executive director of SAG, explained: "It's very effective to use the worker face and the employer face together. It just works. We also have different political lines. Unions have one way, the studios often have another, and that creates goodwill on both sides of the aisle when we do that together" (Interview with authors 2006). The idea that media conglomerates need subsidies and tax relief was, however, a tough sell—even for some members of the coalition. An IATSE activist told the *Nation* magazine, "It's really just about countering Canadian subsidies with our own subsidies and offering even more corporate welfare to the media giants" (Cooper 2000). But the compelling point was the idea that a relatively small investment by government would keep money and jobs in the United States.

After close to five years of lobbying, Congress passed a tax write-off for U.S.-made film and television productions. It appeared in the American Jobs Creation Act, a broad package of tax measures affecting a number of sectors and industries that was signed into law in 2004. Michael Apted, the president of the Directors Guild of America, commented, "With the new incentives provided in today's legislation, uniquely American stories can now be more easily shot in the American towns, cities, and rural areas where they are set—protecting one of our most valuable national resources—our creativity" (DGA 2004). Three years later, the details of how to calculate the production incentive were still being worked out (SAG 2007b). But regardless of the details of the legislation, the alliance's key victory was getting onto the public agenda the idea that the U.S. film industry needed support. This in turn has helped in similar campaigns at the state and local level. According to SAG's Pamm Fair, "We have legislation now in twenty-two states that provides incentives for filmmakers to work in any given state. They range from waiving sales tax for goods and services to very lucrative rebates" (Interview by authors 2006). Some states, such as Illinois and Florida, have established special funds. Others waive hotel taxes for movie productions. Louisiana and New Mexico have among the most generous incentives in the country. Some cities have their own incentives, including San Francisco, which approved a rebate package in 2006 (Tan 2006).

Tied to the campaign for incentives is the idea that film and television are essential components of American culture, not just the U.S. economy. Led by

SAG, the Hollywood unions have combined an appeal to prevent runaway production on economic grounds with an appeal to protect American culture on patriotic grounds, a mix that has resonated particularly well in the years since the attacks of September 11, 2001. This is not the first time the unions have equated foreign production with anti-Americanism. In 1959, when the American Federation of Labor's Film Council called on Congress to launch a full-scale investigation of runaway production, it charged that "in most instances 'runaway' pictures made by American producers in foreign countries give employment to known Communists and thus give aid and comfort to the Communist conspiracy against the free world" (SAG 2007c). The message this time was less crude. It was based on the idea that American stories should be told in America, by Americans. "We've used some public relations in identifying films that are very American stories—*Cold Mountain* was one of them—where we say, 'You may have thought this was an amazing American production but it was done in Romania,'" says Fair. The Film and Television Action Committee (FTAC), an association led by below-the-line unions that includes SAG and some civic and professional organizations and is dedicated to fighting runaway production, launched an e-mail and Internet campaign against the 2003 film. An article posted on its website urged a boycott of the movie, saying, "You can send a message that these economic losses and artistic choices compromised in the name of saving money are not acceptable to Americans" (quoted by Horn 2004). FTAC also called for a boycott of the USA Network's film *Rudy*, the story of New York mayor Rudy Giuliani, because it was filmed in Toronto. The group's statement charged, "This is about patriotism. This is about one of America's darkest hours, 9/11. This is about the American Spirit. How could anyone think it was ethically or morally right to film this story in another country?" (quoted by Elmer and Gasher 2005b, 3). The guild has urged high-profile members to refuse to take part in runaway productions, or at least to speak out against them. At the 2004 SAG awards, for example, actor Tim Robbins implored "the power in this room" to stem the tide of runaway production and to "bring back some of those productions into the United States of America" (Brodesser 2004).

In addition, in 2002 SAG decided to extend its jurisdiction beyond the U.S. border by making its membership Rule One global. The rule, which SAG sees as a founding principle of the union, says: "No member shall work as a performer or make an agreement to work as a performer for any producer who has not executed a basic minimum agreement with the Guild which is in full force and effect" (SAG 2003, 33). Until then, the rule applied only to productions filmed in the United States, not to those made abroad. From May 2002 on, it would apply globally. The Canadian Film and Television Producers Association (CFPTA) denounced the extension. According to a Vancouver law firm active in the film industry, the association had good reason to do so: "The imposition of a SAG agreement on U.S. actors working in Canada can mean a significant increase in pension and welfare payments (especially on a series) and in the residuals that are paid to actors. Further, the producer employing the American actor may be required to become a signatory to the SAG agreement and to offer other performers similar arrangements" (Roberts and Stahl 2002). The association also suggested that the move would undermine the Canadian performers union, the Alliance of Canadian Cinema, Television and Radio Artists or ACTRA (CFTPA 2002). ACTRA responded almost immediately with a statement denying this suggestion, clarifying ACTRA's position on dual unionism, and noting that the Canadian union is proud of its "long-standing positive and mutually supportive relationship" with SAG (ACTRA 2002). Before imposing the change, SAG had consulted with performers' unions in Canada and Australia. In addition, it had lined up the support of key Hollywood unions, including AFTRA and the writers and directors unions, all of which issued statements supporting the move.

A SAG news release, issued on the second anniversary of the Global Rule One campaign, claimed it had been a major success. It said the union had secured 69 foreign theatrical motion pictures in 2002, 120 projects in 2003, and 60 in the first part of 2004. Guild members' earnings on foreign productions covered by SAG contracts topped $120 million, and an additional $6 million from foreign productions went into the union's pension and health funds (SAG 2004). Fair said one of the positive consequences of Global Rule One has been to demonstrate to film workers in other countries the advantages of a SAG contract. This could help raise the bar for other workers and their unions (Interview with authors 2006).

The successes Hollywood unions have enjoyed in the fight against runaway production have had much to do with their ability to form and sustain coalitions, mainly with other unions. The groundbreaking study of the problem was a cooperative effort between SAG and the Directors Guild; the Runaway Production Alliance included a cross-section of the industry; and SAG's move to extend Rule One drew on support from U.S., Canadian, and Australian unions. But while the coalition approach has been effective, coalitions are inherently unstable. As earlier chapters have demonstrated, a number of unions in the communication sector have pursued more permanent arrangements to increase labor power, or present a united front to capital. Would such a strategy work in the fight to protect jobs in the film and video industry? One of SAG's strongest allies in building and maintaining coalitions is AFTRA, which represents seventy thousand performers: sound recording artists; radio and television journalists; performers in soap operas, talk shows, reality shows, and game shows; and film, television, and commercial actors. About forty-four thousand actors hold memberships in both unions. As chapter 5 described, all attempts to merge the two unions have failed, due mainly to a significant minority of SAG members who are not interested in belonging to a union that would include a broader range of performers.

Since the most recent failed merger attempt, in 2003, AFTRA has been rethinking its relationship with SAG, including its continuing support of the runaway production campaign. It has had no interest, for example, in joining the Film and Television Action Committee (FTAC) campaign to put together a complaint to the U.S. trade representative's office about Canadian subsidies. Nor has the Directors Guild, which concluded that a trade war would hurt everyone in the business. SAG, by contrast, has supported the plan. Its board voted in late 2005 to contribute $10,000 to FTAC for a report on the issue (McNary 2005). It reiterated that support a year later with a promise of a $50,000 payment to the law firm putting together the initial filing of the complaint (SAG 2006). FTAC has argued that filing a complaint alleging that Canadian film production subsidies violate trade agreements is the best way to combat runaway production. When the idea was proposed several years ago, Magder and Burston argued that it would be a lengthy and costly venture with relatively little chance of success (2001, 223–24). That prediction proved prescient. In the fall of 2007, U.S. trade representative Susan Schwab announced that she would not launch a formal probe into the legality of Canadian film and TV subsidies. She found that neither the economic data nor the legal arguments set out in the petition supported FTAC's contention that film subsidies are unfair (*CBC News* 2007).

This section has described how the Screen Actors Guild and other Hollywood unions have identified and then taken action to fight a form of offshore outsourcing that has cost American film and video workers jobs. SAG has commissioned reports, lobbied for legislation, launched publicity campaigns, experimented with expanded jurisdiction, and worked with FTAC on the ultimately unsuccessful trade complaint. In all of these measures, it has worked with other organizations, mainly trade unions representing both above-the-line creative workers and below-the-line craft and technical workers, but community and industry groups as well. Some of these efforts have had a positive effect for labor. Perhaps the most significant is that the film industry coalitions have managed to put the issue on the U.S. political agenda, at both the national and state level. The federal government and more than twenty states now have film incentive legislation or tax rebates of some sort. Whether they are strong enough on their own to keep production in the United States is difficult to say. Ironically, their appearance coincided with a rise in the value of the Canadian dollar, a development that in its own right made Canada less attractive as a production site. The Canadian dollar, which hit a record low of just under US$0.62 in 2002, was roughly on par with its American counterpart by the end of 2007. In August of that year, Canadian film executives in British Columbia called for an industry-wide agreement to discount the dollar for foreign investors. In essence, the industry would act as if the Canadian dollar were worth US$0.88 (Webb 2007). Moreover, other countries, such as Romania, are happy to offer themselves to Hollywood as low-cost production sites. The decision to make SAG's Rule One global has put more money into SAG's coffers. The appeals to patriotism and the threats to boycott productions filmed abroad may be having a psychological effect. Pamm Fair of SAG said that when Hollywood film and television producers, and even independent producers, do their budget these days—"and we've seen some of them"—they compare costs for Canada and two or three other places in the United States, such as New Mexico, Louisiana, Los Angeles, or New York. "If the gap is narrow enough they'll do it in the U.S.," she said (Interview with authors 2006).

Although they have been effective, informal coalitions work only as long as they meet the self-interest of individual members. As a rule, their focus tends to be narrow, driven by a single issue—or a small collection of related issues—on which the various members may agree. In addition, alliances come and go. As the relationship between SAG and AFTRA soured in 2007, relations between SAG and the Writers Guild grew warmer. The ever-shifting state of an informal alliance means that unions that should be allies may find themselves in competition with each other, knowing all the time that the real adversary is the conglomerates that dominate the industry, not the other unions. The fight against runaway production raises another question about worker solidarity. Even if the Hollywood unions were able to merge and better defend their members' interests, what would this mean for film workers in other countries? Don't these workers have every right to the cultural industry jobs that SAG, AFTRA, and other unions want to keep in the United States? AFTRA national recording secretary Matt Kimbrough said his union is aware of the problem. "We take the position that it should not be a worker-against-worker competition. The only way you remove that incentive is to raise the boats in other places. It's not going to help us to go after the work that other unions in other countries do. We need to support all unions and the rights and wages of all the workers in all the countries" (Interview with authors 2006).

Building an International Labor Network

As important as it is for cultural workers of the United States to unite, it is far more important, and arguably essential to the labor movement, for knowledge and cultural workers to unite around the world. This section explores that possibility by describing evidence of resistance to outsourcing, organized efforts to advance this opposition, and the attempts by one global union federation to build international worker convergence.

Just as offshore outsourcing is growing, so is resistance to it from outside the developed world. One of the more noteworthy signs of this resistance is based in India: the New Trade Union Initiative, which brings together about one hundred labor unions across that country. In December 2004, the Initiative sent a delegation to the United States to meet with trade union and worker association leaders in nine cities to discuss common strategies for dealing with outsourcing. This is an important development because it is the first step toward bridging the fundamental divide between workers in nations that are losing jobs and those in nations that are gaining jobs. Indian trade unions support the creation of new jobs in the high-tech and services sectors, but they are concerned about working conditions and the sustainability of such jobs. As one visiting labor leader put it,

> Jobs are going to India not because of the wage difference but because these jobs are unregulated. There are no laws in India about minimum wage or the maximum number of hours workers can work. And multinational companies are taking advantage of that . . . People are working 16-hour days and often nights, at five times the intensity of the American workers who do the same job. (Tollefson 2005)

Not unreasonably, Indian workers who see the growth of a high-tech presence in China and elsewhere in Asia, often facilitated by Indian firms, worry about just how long the high-tech boom in India will last. As another put it, "We now have global mobility of capital without global mobility of labor . . . Why not build a global framework of contracting?" (Tollefson 2005). In a sign that Indian resistance is having some impact, the *New India Express*, a leading English-language daily, published an editorial in 2005 calling for legislating an end to trade unions and strikes. Following a general strike that saw 60 million workers walk off the job, the newspaper declared, "Perhaps the time has come for the Congress to get its act together and actually implement some neo-liberal policies. It could begin by banning trade unions and strikes and freeing India of this scourge forever" (*New India Express* 2005). Workers' organizations in the developed and developing world know that building a global framework for labor will not be easy. Interestingly, there are signs that China may force the issue, as labor unrest there is on the rise. Workers in China staged an astounding fifty-seven thousand strikes and protests in 2004. Moreover, labor reform training centers have opened up across the country and cadres of knowledge workers, trained in the law and experienced in the workplace, are carrying out worker training and arguing cases before the courts (Roberts 2005; Zhao and Duffy 2007).

Organizing has also taken place through initiatives that bring together international labor federations and workers in places that have been major recipients of outsourced jobs. For example, the Union Network International (UNI), a major global labor federation, has supported the work of the IT Professionals Forum—India in its efforts to protect the interests of workers in the business processing outsourcing sector from unjust labor practices. Public meetings and organizing drives have taken place in Bangalore and Hyderabad. Like WashTech in the United States, these groups style themselves more along the lines of guilds or social movement workers' organizations than traditional trade unions. As a result—and again like WashTech—they see themselves as better able to represent workers in the information-technology sector.

Although UNI is a relatively recent creation, founded in 2000, developments such as its Indian initiative make it worth looking at closely. The rest of this chapter examines this global federation in depth. It is worth noting at the outset, however, that UNI is not the only international labor organization born recently from convergence. Rather, it is part of a growing pattern of globalization among trade unions. Although this phenomenon is occurring primarily at the level of international federations, we are also seeing global labor convergence among some individual unions as well. One significant example is the 2007 decision by the UK union Amicus and the United Steelworkers union from the United States to take the next step toward a trans-Atlantic merger.

As a global federation that spans all sectors of the converging electronic services arena, UNI is situated between two very different types of global federation. On the one hand, there is the International Trade Union Confederation (ITUC), which was created in November 2006 when delegates from 156 countries met in Vienna for an event unprecedented in the history of the international trade union movement. They dissolved two globally operating and competing international confederations—the International Confederation of Free Trade Unions and the denominationally oriented World Confederation of Labor (WCL)—and founded the ITUC, comprising 304 affiliated federations in 156 countries in which 168 million workers are organized across the world. According to one recent analysis, "Congress delegates were firmly convinced that the globalization of political institutions and the globalization of business and markets must be followed by the globalization of trade unions" (Traub-Merz and Eckl 2007, 1). ITUC represents the broadest level of convergence at the international level, spanning all industries and all types of unions. Because it formally represents trade unionists worldwide, it holds considerable potential for the labor movement, including the promise of unprecedented global coordination of labor's response to transnational capital and neo-liberal governments. But it is too early to determine whether the federation will live up to its promise, or get bogged down in more mundane bureaucratic and political disputes.

On the other hand, there is the International Federation of Journalists (IFJ), an example of global labor convergence that focuses on one sector of the media industry. The IFJ is the world's largest journalism organization, representing 500,000 journalism professionals who comprise its 161 member unions from 117 countries. It has been around for a long time: established in 1926, relaunched in 1946, and reorganized into its present form in 1952. Its mandate is to promote international action to defend press freedom and social justice through strong, free, and independent trade unions of journalists. By focusing on one industry, the federation has been able to concentrate on specific problems facing media workers, such as violence against journalists, media ownership concentration, women's rights in the media, authors' rights to control their work, and institutional attacks on press freedom. The IFJ also aims to bring together journalists from both rich and poor nations to develop a common platform to address outsourcing media labor. Because it is focused on one industry and has a clearly defined set of goals, the IFJ has been able to avoid the bureaucratic problems and political disputes over priorities that often beset larger federations. But as technological and corporate convergence challenge traditional definitions of journalism—and as some of its member unions, like the CEP in Canada and the CWA in the United States, enlist workers across both the content and technical segments of the knowledge industries—questions arise about whether the IFJ can continue to succeed by focusing on one media sector.

UNI sits somewhere between the breadth of ITUC and the specificity of the IFJ. Unlike the IFJ, UNI fully embraces convergence, calling itself "a new international for a new millennium." Unlike ITUC, it does not seek to represent all things to all types of workers. It primarily spans the newly converged electronic information and communication sectors, including workers in the postal, media, entertainment, telecommunications, and cultural industries. As with other labor convergence projects, a driving force behind its creation was the growth of companies that span these sectors by taking advantage of converging electronic technologies. Although it is still a very young organization, UNI has been in the forefront of global labor issues like outsourcing, and prominent in applying pressure to global companies and global organizations like the World Trade Organization. From its inception, a central focus of UNI has been to make inroads in the high-tech and service sectors of the economy (Morgan 2000).

UNI was formed on January 1, 2000, through a merger of four global unions: Communications International; FIET (the white collar and services grouping); the International Graphical Federation; and the Media and Entertainment International. General-Secretary Philip Jennings used the language of convergence in explaining the need for such a global organization, saying, "Workers across the service, communications, and entertainment sector face the same employer and the same issues and are increasingly lobbying the same international organizations" (quoted in International Centre for Trade Unions Rights [ICTUR] 2005, 408). If workers around the world work for the same multinational employers, the most effective strategy, UNI contends, is to coordinate and act collectively on a global scale by forming global union networks to confront these multinationals. UNI is affiliated with other global union federations and is funded by fees collected from affiliated unions across the globe.

As a result of the merger and of other unions choosing to affiliate on their own, UNI comprises 15.5 million workers in 900 unions in 140 countries. It is divided into twelve general economic sectors, each of which has autonomy in how it chooses its president, vice presidents, and committees, and in the determination of policies, priorities, and structures. For example, the media, entertainment, and arts sector deals with issues such as media ownership and concentration, the structural evolution of the sector, trade policies, legal standards, technological change, copyright, and censorship (U.S. Newswire 2005). The UNI telecom sector handles liberalization, deregulation, and privatization agendas and mergers and acquisitions that are driving the restructuring of global telecommunications markets and policies (UNI 2004b). Industry-wide coordination on a global scale is seen as an effective way to build real worker power and is the underlying rationale for establishing these sectoral groups (U.S. Newswire 2005). Although the aim of the federation is to unite workers across regions, it nevertheless finds it valuable to divide its activities into four broad regional groupings: Europe, Africa, Asia-Pacific, and the Americas. In many countries, there are also national liaison councils. Each region is responsible for assisting in the development of independent and democratic unions, and each operates under its own statutes and standing orders.

UNI also has special committees to address issues that cut across all sectors, including youth, women, and professional and managerial staff. The federation recognizes that some of its efforts, such as educational campaigns, need to be directed toward integrating youth into the trade union movement so that they become active and informed members. Young people are often in a more precarious employment relationship than older workers, and because of inexperience they are more vulnerable to employer exploitation. Each of the four regions has its own youth section to deal with issues specific to that geographic area. The UNI women's committee recognizes that no matter where they work or how old they are, women are in a more precarious employment relationship than men due to sexism and patriarchy that exist across the globe. Violence against women in the workplace is still a common occurrence in many parts of the world. In addition, working women everywhere face the double day, because most family responsibilities still fall on them. Getting global employers to institute family-friendly policies, such as adequate maternity leave, is one of the federation's major campaigns. The UNI women's committee helps working women deal with these issues. In addition, it seeks to recruit more women into the union movement so that they too become active and informed members (UNI 2004a, 2005i). The UNI professional and managerial committee deals with issues facing highly qualified workers in all sectors that UNI covers. It recognizes that while workers like engineers and scientists are often referred to as professional or managerial staff, not all of them occupy actual managerial roles. Job content and qualification levels may distinguish them from other categories of workers, but as waged workers they still need organizational protection from their employers. Because these workers often travel across borders to work on either a short- or long-term basis, UNI has developed its own "passport." The union claims it "helps ensure that you take your access to trade unions with you when you move abroad to work" (UNI 2005f).

UNI does not explicitly aim to stop economic globalization, but rather to give it a social or human dimension. In other words, it does not oppose the global spread of business but maintains that the needs, interests, and rights of workers and the communities and environment they live in cannot be sacrificed in this process (Adler 2001). Jennings notes "the race to the bottom—where the globe is scoured for the cheapest and most exploitable labor—should be replaced by an enlightened race to the top" (UNI n.d.-a). This is, in essence, a reformist position rather than a radical one, and it has put UNI into conflict with some of the more militant trade unionists. At the same time, one of UNI's major aims is to establish rights for workers worldwide. It also contends that unions, like corporations, must become global players and take an active part in shaping the direction of the global economy. Unions need to engage in constructive dialogue on a global scale with business, government, and other international bodies so that labor can have a voice in directing the global economy (UNI 2004c).

To this end, UNI has favored the negotiation of Global Framework Agreements (GFAs) with multinational corporations. GFAs are intended to establish labor and union rights for workers regardless of where a corporation may operate (UNI 2007c). These agreements are based on the International Labour Organization's (ILO) *Declaration on Fundamental Principles and Rights at Work* (ILO 2002). For multinational corporations, signing a GFA can enhance their reputation by demonstrating some concern for corporate social responsibility. UNI sees these agreements as the most effective way to ensure decent labor standards worldwide. More than a dozen corporations have signed a GFA with UNI. While certainly not perfect in terms of protecting and establishing workers' rights, these agreements can be seen as a significant success for the federation. At its Second World Congress, held in Chicago in August 2005, UNI announced that pressuring multinational employers to sign GFAs would be one of its main objectives in the coming years. It hopes to focus these activities on a list of one hundred multinational employers, highlighting Wal-Mart in particular (Marquez 2006).

If UNI is going to be the new international it proclaims itself to be, getting corporations to sign and abide by a GFA will be a mark of its success. GFAs are intended to establish the groundwork for the dialogue and relationship between a multinational employer and a global trade union federation with regards to a corporation's global activities (Waterman and Timms 2004/2005). UNI sums up the strategy as follows: "When companies were local, unions had local agreements; when companies were national, unions had national agreements. Now in the global economy we need global agreements" (UNI 2007c). GFAs are not a substitute for direct negotiations between a corporation and workers at a local or national level. Wages, benefits, and terms and conditions of employment are still established through collective bargaining. But the GFA is to be used as a basis for these negotiations (UNI 2007c). In this sense, UNI operates like an international federation, not a global union. In other words, it transcends national boundaries; the national union of the respective country maintains autonomy in direct negotiations with the corporation that is operating within its jurisdiction.

Two notable exemplars of the GFA process involve a telecommunications firm and one of the world's largest media companies. The first concerns Telefónica, a global telecommunications corporation operating in Spain, Portugal, Romania, Argentina, Brazil, Chile, Peru, Mexico, and the United States. Negotiations began in April 2000 and the agreement was reached in March 2001. According to the ILO, a UN agency, the agreement between UNI and Telefónica represented "a historical milestone in industrial relations" (ILO 2002, 128). Under the agreement, Telefónica and it subsidiaries and affiliates pledge to respect the environment and trade union rights, and establish minimum labor standards. The agreement includes provisions such as a ban on child labor, no employment discrimination, a minimum wage standard, regulated work hours, stable employment, training opportunities, and health and safety standards for the 120,000 workers in the nine countries where it operates (ILO 2002; Sweeny 2001). The deal was binding for five years, after which it was subject to review.

Just because Telefónica signed a GFA with UNI does not guarantee that Telefónica and, more particularly, its subsidiaries and affiliates will abide by it, however. Indeed it is especially difficult to enforce the GFA at the company's many subsidiaries. For example, in Brazil, the UNI-affiliated Sindicato dos Trabalhadores em Telecomunicações (SINTETEL) signed a collective agreement with the Telefónica subsidiary Atento. SINTETEL noted that the GFA was paramount in its ability to successfully launch the campaign (Howell 2006; UNI 2006b). At another Telefónica subsidiary, Atento Puerto Rico, it was a different story. Management of the call center undertook antiunion activity, and nineteen organizers were fired for union activities, though ultimately the union drive was unsuccessful (UNI 2006e). In Puerto Rico, where U.S. labor laws apply and where an antiunion managerial mindset is widespread, the UNI-affiliated CWA met considerable resistance from local Atento management, which violated both the GFA and U.S. National Labor Relations Board procedures.

The president of the UNI Telecom sector, Neil Anderson, went to Puerto Rico in October 2006 to discuss the matter with workers so that he could compile a report and make recommendations to Telefónica on how to avoid similar problems in the future. He expressed outrage with the way that management at Atento dealt with the drive:

> This serious violation of the UNI-Telefónica Agreement must never be allowed to happen again and Atento must make some gesture to show they will respect the agreement in future organizing efforts by the CWA . . . We cannot allow one "rogue" company to destroy the credibility of an agreement that gives workers rights and decent jobs and is acknowledged by both unions and managers as an excellent tool for good industrial relations. (UNI 2006e, n.p.)

Ultimately, the U.S. National Labor Relations Board forced Atento to post notices in the workplace stating that it would not violate labor relations laws in future unionization attempts. The UNI Telecom Global World Meeting in June 2006 passed a resolution condemning the actions of Telefónica and Atento.

The Telefónica GFA expired in 2006 and as of early 2008 was still being reviewed and revised by UNI and Telefónica. Since one of the main stipulations of the GFA is to allow for union organizing activity to occur, a test of the feasibility of the entire GFA mechanism awaits another, similar unionization attempt and a demonstration of UNI's ability to deal with it. Indeed, this could have an impact on the success or failure of UNI. Meanwhile, UNI is not backing down. The unionization drive at Atento Puerto Rico remains a fundamental issue in the ongoing discussions and negotiations with Telefónica (UNI 2006c, 2006e). At the UNI Global Union Telefónica Alliance meeting in Chile in January 2007, attended by top Telefónica management, UNI reiterated that the GFA needs to be respected wherever Telefónica operates—and particularly in the subsidiary companies. UNI also called for Telefónica to reduce the levels of outsourcing and precarious employment. The federation also emphasized that Telefónica needed to retain and further develop a skilled workforce in the telecommunications, media, Internet, and information technology-enabled services industries so that these technologies and services can be properly produced and maintained. Telefónica management responded that outsourcing is the only way for the company to stay competitive, but agreed on the need for a skilled workforce. It also pledged to work harder to ensure company-wide adherence to the GFA (UNI 2007a).

In April 2007, UNI signed a GFA with the global media giant Quebecor World, which employs 35,000 workers in more than 160 facilities in the United States, Canada, Brazil, France, the United Kingdom, Belgium, Spain, Austria, Sweden, Switzerland, Finland, Chile, Argentina, Peru, Colombia, Mexico, and India. In a "Joint Statement on the Respect and Promotion of International Labor Standards," Quebecor World committed to ensuring that fundamental labor rights, derived from ILO Conventions, will be respected. Specifically, the company affirmed that it will not use forced, bonded, or child labor; that there will be no employment discrimination; and that all workers will have equal opportunity and receive equal pay for work of equal value. Quebecor World also pledged that it will not prevent workers from organizing and will not dismiss or discriminate against any employees because of union membership or participation in unionizing activities. It also committed to respect the right of unions to represent employees and to bargain collectively on employment terms and conditions. Quebecor World has agreed that it will provide its employees with a clean, safe, and healthy place to work and will develop policies aimed at creating respect in the workplace. The company and UNI have committed to ongoing cooperation and dialogue to ensure their mutual interests as well as to resolve any differences and disputes (UNI 2007b).

As a global union, UNI finds the international outsourcing of jobs, especially in the telecommunications industry, to be one of its major challenges. The federation has affiliates in countries that gain jobs through outsourcing, and in countries that lose jobs through outsourcing. Since it does not want to be accused of being racist, xenophobic, or protectionist, the union does not oppose outsourcing in principle. Rather, it fights to stop unconstrained and unregulated forms of outsourcing—such as outsourcing to export processing zones where neither labor nor environmental standards exist and outsourcing that directly leads to layoffs in the country from which the work was sent (UNI 2004c, 2005a). In a position paper on the issue, UNI's Gerhard Rhode contends,

> While it is understandable in rational and geopolitical terms that the outsourcing countries want to fight poverty and struggle to catch up with the world-

> wide economy, and that it is important to create jobs in developing countries, a union cannot treat it as an abstract consequence when its own members lose their jobs. Job security is a central concern for UNI, and if work is going to be outsourced, consultation with unions should occur, displaced workers should be retrained, and there should be help for them to find other comparable work .(Rhode 2003, n.p.)

The second part of UNI's strategy for dealing with outsourcing is to promote and support unionization drives in countries where outsourced work is done. This is seen as a way to establish decent labor standards, similar to those stipulated in the Telefónica GFA (UNI 2004c, 2005a). The hope is that organizing workers in countries that receive outsourced jobs can raise labor standards globally (Marquez 2006). UNI has been criticized for not being sufficiently mindful of workers in developing countries (Moore 2005). UNI disagrees, stating that its regional committees have substantial autonomy in how they go about organizing and recruiting workers as well as in setting priorities. Considerable energy is directed toward projects for workers in outsourced call centers and in back offices in developing countries such as India. Call centers are a special focus for UNI because they cut across industrial sectors, are growing quickly, employ large numbers of younger workers and large numbers of women, and often have less than favorable working conditions (UNI n.d.-b and c).

UNI is unequivocal in its support for global labor solidarity and, unlike some other federations, for encouraging new forms of worker associations. As Andrew Bibby noted in a position paper for the federation:

> Reinforcing international trade unionism [includes] encouraging practical solidarity between unions in developed countries and those in developing countries [and] working to welcome embryonic forms of trade unionism in new industries and new countries, recognizing that they may be different from traditional union structures. (Bibby 2004)

Global solidarity campaigns, support work for national unions facing converging industries, and alliance building are major components of UNI's work. Specifically, this includes supporting union recruitment, organizing workplaces, and collective bargaining; undertaking campaigns to bring global attention to specific labor issues; holding sectoral, regional, and global conferences to deal with labor issues; and participating in social activism.

Getting workers in outsourced call centers and other information technology businesses to support unionization is difficult. An editorial in the *Australian* contends these workers "think of themselves as members of a relatively well-paid, respected professional elite in no need of union protection"

(2005). In this respect, they are much like the high-tech workers on the U.S. west coast, who formed WashTech despite significant discomfort among potential members. UNI recently launched IT Professional Forums in India, in an attempt to reach out to nonunionized workers. The term "forums" is used because of concerns that using a term like "trade union" might discourage potential members who have no trade union experience. UNI hopes that these types of organizations will win over Indian workers who may have some misgivings about trade unionism (Bibby 2002, 2004; Srinivasan 2005). However, trade union consciousness is not absent. As a worker at one the forums commented, "Unity is strength—and when capitalists get collectively organized, why shouldn't we?" (quoted in Bibby 2002). These forums are helping workers to defend their rights, including against unfair managerial practices (*Financial Express* 2004).

A related UNI project, but one with a more traditional trade unionist orientation than the forums, is the Centre for Business Process Outsourcing Professionals that opened in Bangalore in August 2004. This led in 2005 to the formation of the Union for Information Technology Enabled Services Professionals (UNITES), a trade union that is registered with the All India Trade Union Congress. UNITES hoped to increase its membership from nine thousand to twenty thousand by the end of 2007 (Mishra 2007). In May 2006, the union signed its first collective agreement with Excel Outsourcing Services covering sixty-five workers. UNITES sees its success in completing this collective bargaining agreement as modest, but UNI views it as a major breakthrough and continues to provide UNITES with financial and other forms of support (UNI 2006f). UNI's support is especially important in India because there is considerable pressure to restrict union activity, especially in the information-technology sector. It is also increasingly essential because of growing reports of significant health issues arising from the stress of working in the information-technology industry (Mahapatra 2007).

One of UNI's most significant campaigns has been directed at reform of the global call-center industry. To that end, it set forth a Call Centre Charter and produced newsletters and advisories on global, regional, national, and local challenges, problems, and activities specific to call-center workers (UNI 2005d, n.d.-b, n.d.-c). The Call Centre Charter sets out minimum standards for call centers worldwide and is intended to act as a guide for employers. It establishes standards for health and safety, hours and workload, surveillance and privacy, pay and benefits, training and promotional opportunities, and the right to organize and bargain collectively. The newsletters report on such things as campaigns, actions, conferences, and organizing attempts, defeats, and successes. One example is a union organizing and information seminar for call-center workers in Brazil, attended by one hundred young women. The seminar was meant to develop strategies to empower workers on the shop floor. Within ten days of the conference, these women went back to their respective call centers and signed up six hundred workers for membership with the UNI-affiliated SINTETEL union (UNI 2004d).

Support and solidarity work is another major aspect of UNI's activity, and much of this is done in conjunction with other global union federations. Solidarity work includes raising global awareness of labor issues in particular parts of the world and writing letters of protest to governments and the upper echelons of management of global corporations. UNI also undertakes corporate campaigns, which include highlighting the misdeeds of particular corporations and generating public pressure on corporations to act in a socially responsible manner. Part of the plan is to expose corporate bad behavior through naming and shaming campaigns (Ragan 2003). But if solidarity work is going to be more than just posting articles and bulletins on the web, UNI needs to demonstrate what positive changes this work has brought for workers who requested solidarity. Boasting, as UNI does, that it will respond to a call for solidarity assistance within twenty-four hours is commendable, but whether the response leads to significant action is another matter (Jennings 2006). The federation has demonstrated a willingness to take innovative approaches to its solidarity campaigns. For example, in 2007 it joined with nine thousand workers striking against IBM in Italy by helping to launch a global virtual strike in the popular 3D social networking site Second Life. UNI saw the action as more than just workers' avatars carrying picket signs. It was also an opportunity to build solidarity with workers in eighteen countries who joined the online action in support of their fellow unionists in Italy (Smith 2007).

UNI also supports the concept of social movement unionism as part of its strategy to expand trade union influence and relevance to workers, especially in developing countries. It sees this as a way to supplement traditional trade unionism with other forms of worker and community empowerment, and to establish connections with progressive elements of civil society including cooperative movements and mutual aid societies (UNI 2003a, 2006d). UNI is also involved in global movements such as the Make Poverty History campaign, focusing specifically on trade injustice throughout the world, and the global fight against HIV/AIDS (UNI 2005b, 2005c).

UNI believes that strategically coordinated global union action is the only way to address the problems posed by multinational corporations. While some national unions may be riding UNI's coattails as a way of claiming to be involved in the global labor solidarity movement, the organization is more than simply a network of affiliated unions. UNI has a fairly well-thought-out plan of action. There is evidence of support from unions in advanced capitalist nations for unionizing activities and the development of other forms of worker organization. Since UNI is only seven years old, its full potential is still uncertain. However, it is clear that the federation represents the global extension of a form of convergence that has received all too little attention.

Conclusion: The Laboring of Communication

We began this book by asking a simple but provocative question: Will knowledge workers of the world unite? Earlier chapters have shown that a significant degree of unity is not only possible but is actively being pursued within and across national borders in Canada and the United States. North American unions have pursued trade union mergers and bargaining-unit consolidations. They have attempted to build bridges between labor and social movements, and to create new forms of worker organization for those who cannot or will not join a traditional union. Nonetheless, trade union density continues to decline, and organized labor is struggling to retain existing membership in the industrial economy at the same time as it tries to expand membership in the postindustrial economy.

This chapter has looked at whether or how labor unity can be pursued at the international level, by focusing on the critical problem of outsourcing knowledge work. While it is uncertain how rapidly the practice of outsourcing will grow over the coming years, and in precisely which configurations, there is no doubt that it will be a defining issue for knowledge workers and their organizations in all parts of the globe. The bulk of the evidence suggests that outsourcing is a major step in the deepening and extension of global capitalism. Outsourcing is also multifaceted—one vector of an increasingly complex international division of labor, involving far more than the transfer of media or information service jobs from high-wage nations to low-wage nations. Most outsourcing occurs within the developed world, but the proportion of outsourcing going to developing nations is increasing rapidly. At the same time, we are seeing the rise of outsourcing within and among the less developed countries, as well as some that originates in countries like India, where companies are launching satellite operations into the heartland of Western capitalism. The accelerating movements of labor worldwide follow the circulation of capital in ways that challenge simple notions of dominant and dependent nations. Deep inequalities persist among nations, but even those suffering some of the most extreme consequences of poverty and colonialism contain players participating at the leading edge of global capitalism.

Western workers both resist and welcome outsourcing, depending on whether they are losing or gaining jobs. Workers in the less developed world tend to welcome outsourcing because it tends to make jobs available, but are less supportive when—as is often the case—these jobs contain none of the regulatory protections that limit hours, set a minimum wage, and provide adequate occupational safety and health standards. The result is a flurry of new forms of workplace resistance that challenge traditional trade unions and give rise to new forms of labor organization, as well as to new forms of global labor collaboration. Since this activity increasingly involves the knowledge, information, and media sectors, scholars who study these areas would benefit by paying closer attention to these new forms of what Denning (1996) called "the laboring of culture."

Outsourcing is both enabled and promoted by the same sorts of factors that have put trade unions in North America on the defensive. These include the rise of a neo-liberal economic environment that favors a business agenda, regardless of the consequences for employment and social welfare; the development of information technology that permits companies to deskill or dismiss workers and contract out as many functions as possible; and a steady increase in corporate concentration, not just within the home country but at the international level as well. But as this chapter has shown, convergence also applies to movements that resist outsourcing, including established North American–based trade unions, new forms of worker organizations in developed and developing countries, and new networks, assisted and encouraged by global labor federations, to link workers. This resistance makes prediction much less certain, but it helps explain the complex dynamics of outsourcing and the policies and practices workers are developing to deal with it.

One way trade unions can make a difference is by working together on a particular issue. As the case study on runaway production demonstrates, coalition building can be effective, especially when the coalition can draw on a diverse range of groups with a stake in the issue at hand. The unions managed to get the question of runaway production onto the congressional agenda. In addition, the Screen Actors Guild has succeeded to a considerable degree in making its Rule One global. At the same time, however, the fight against runaway production is clearly protectionist, aimed at safeguarding "American" jobs for American creative workers. This is an increasingly troubling idea in an era when changes in the global division of labor support a smoothing out of cultural differences. Film and television productions are aimed at a global audience, not just an American one, and enforcing Rule One globally will do nothing to curtail this process. And in the long run, protectionism may end up pitting worker against worker. The real task, as AFTRA's Matt Kimbrough puts it, is to find ways to support all unions, and the rights and wages of all workers in all countries.

The Union Network International attempts to do just that, and in a variety of ways. At the most basic is an attempt to bring knowledge workers of all sorts to the same table, a place where they work together and learn from each other. In addition, UNI pays special attention to the needs of some of the most precarious workers: young people, who are more likely to be exploited due to inexperience; women, who are often exploited in both the workplace and the home; and professionals, who think unions are for blue-collar workers, not white-collar workers, but who would benefit enormously from working together on workplace issues. UNI's most innovative tool is the creation of Global Framework Agreements with employers, which set out some fundamental conditions of work and form the groundwork for collective bargaining at the national or local level. These agreements offer the potential of improving basic workplace standards for all workers, and at the same time respecting local differences. All of these actions are limited, and none of them offers a guarantee of global labor solidarity. But they represent a start toward bridging the gulf between the haves and the have-nots in the growing cadres of knowledge workers. It will be fascinating to watch how this organization develops in the coming years.

In conclusion, for too long the concerns of communication and information workers, and the trade unions and social movements that represent them, have languished on the fringes of academic research. It is impossible to escape the irony in this, given that academic researchers are themselves knowledge workers. The lucky ones have secure positions, often in unionized environments. An increasing number, however, work precariously, moving from contract to contract and aware that the only permanent feature of their work is its temporary or contingent nature. Critical communication scholars have devoted considerable attention to understanding how the information society can be a source of liberation for knowledge workers, not just a means of deepening and extending capital's control. If the information society is to be a genuinely democratic one, then it is time we all paid serious attention to the laboring of communication.

References

AAUP. *See* American Association of University Professors.

ACTRA. *See* Alliance of Canadian Cinema, Television and Radio Artists.

Adler, M. 2001. German president calls for internationalization of labor movement. *Agence France-Presse*, January 5. http://global.factiva.com/ (accessed May 15, 2006).

AFL-CIO. *See* American Federation of Labor and Congress of Industrial Organizations.

AFTRA Magazine. 2003. AFTRA and SAG boards approve consolidation plan. Winter.

——. 2004–2005. National board begins AFTRA restructuring. Fall/Winter.

——. 2006. AFTRA on the move. Winter.

AFTRA. *See* American Federation of Television and Radio Artists.

Akyeampong, E. B. 2004. The union movement in transition. *Perspectives on Labour and Income* 5 (8): 5–13.

——. 2005. Business support services. *Perspectives on Labour and Income* 6 (5): 5–9.

Alliance of Canadian Cinema, Television and Radio Artists. 2002. ACTRA statement respecting SAG's Global Rule One. April 17.http://www.actra.ca/actra/control/hotnews1?category=hn&id=10127 (accessed March 20, 2007).

Amber, A., Canadian director of the Communications Workers of America. 2007. Interview with C. McKercher. Ottawa, April 24.

American Association of University Professors. 2007. Contingent faculty. http://www.aaup.org/AAUP/issues/ (accessed September 13, 2007).

American Federation of Labor and Congress of Industrial Organizations. 2005. AFL-CIO Executive Council announces formation of first industry coordinating committee covering the art, entertainment, media and telecommunications industries. October 6. http://www.aflcio.org/mediacenter/prsptm/pr10062005.cfm (accessed August 27, 2007).

American Federation of Television and Radio Artists. 2004. Historic federal runaway bill signed into law. http://www.aftra.org/legislative/docs/federal_runaway_bill_2004Oct.htm (accessed March 20, 2007).

______. 2006. AFTRA introduces national president's message podcast on website. News release, January 4. http://www.aftra.org/press/press.htm (accessed March 16, 2006).
——. 2008. AFTRA national board of directors resolution regarding phase one negotiations. February 2. Available at http://www.aftra.org/ (accessed February 10, 2008).
Anderson, J. C. 1989. The structure of collective bargaining. In *Union-management relations in Canada*, 2nd ed., ed. J. Anderson, M. Gunderson, and A. Ponak, 209–34. Don Mills, ON: Addison-Wesley.
Artz, L., and Y. R. Kamalipour, eds. 2003. *The globalization of corporate media hegemony*. Albany: State University of New York Press.
Associated Press. 2007a. Local news web site postpones coverage by reporters in India. *International Herald Tribune*, May 14. http://www.iht.com (accessed December 21, 2007).
——. 2007b. E.D.S. offers exit incentives to 12,000 workers. *New York Times*, September 14. http://www.nytimes.com/2007/09/13/technology/13data.html?_r=1&oref=slogin (accessed January 4, 2008).
AT&T. 2008. Company information. http://www.att.com/gen/landing-pages?pid=3309 (accessed January 15, 2008).
Austen, I. 2004. Canada, the closer country for outsourcing work. *New York Times*, November 30.
Australian. 2005. 'Cyber coolies' baulk at joining trade union. October 4. http://global.factiva.com/ (accessed June 3, 2006).
Babbage, C. 1832. *On the economy of machinery and manufactures*. Philadelphia: Carey & Lea.
Babe, R. E. 1996. Convergence and the new technologies. In *The cultural industries in Canada*, ed. M. Dorland, 283–307. Toronto: James Lorimer & Company.
Bahr, M. 1998. *From the telegraph to the Internet*. Washington, D.C.: National Press Books.
Baker, E. F. 1974. *Printers and technology: A history of the International Printing Pressmen and Assistants' Union*. Westport, CT: Greenwood Publishing Group.
Balka, E. 2002. The invisibility of the everyday: New technology and women's work. In *Sex and money: Feminism and political economy in the media*, ed. E. Meehan and E. Riordan, 60–74. Minneapolis: University of Minnesota Press.
Balka, E., and R. Smith. 2000. *Women, work, and computerization*. Boston: Kluwer Academic Publishers.
Barley, S. R., and G. Kunda. 2004. *Gurus, hired guns, and warm bodies: Itinerant experts in a knowledge economy*. Princeton, NJ: Princeton University Press.
Barnes, B. 2007. In Hollywood, a sacred cow lands on the contract table. *New York Times*, August 5.
Barnett, G. E. 1926. *Chapters on machinery and labor*. Cambridge, MA: Harvard University Press.
Barr, R. 2005. U.S. tech workers suffer wage loss from offshoring. *Arizona Republic*, September 5. http://www.azcentral.com/arizonarepublic/business/articles/0905techpay05.html (accessed January 1, 2008).
Barrett, T. 2005. To censor pro-union web site, Telus blocked 766 others: Company's labor dispute tactic may affect Internet speech law. *The Tyee*, August 4. http://www.thetyee.ca/News/2005/08/04/TelusCensor (accessed June 5, 2006).

Batstone, E. 1984. *Working order: Workplace industrial relations over two decades*. Oxford: Basic Blackwell.

Beasley, R. E., E. Lomo-David, and V. R. Seubert. 2001. Telework and gender: implications for the management of information professionals. *Industrial Management & Data Systems* 101 (8/9): 477–82.

Becker, A. 2007. Old media, new media. *Broadcasting & Cable*, February 26.

Bell Canada Enterprises. 2008. About BCE. http://www.bce.ca/en/aboutbce/ (accessed January 15, 2008).

Bell, D. 1973. *The coming of post-industrial society: A venture in social forecasting*. New York: Basic Books.

——. 1976. *The cultural contradictions of capitalism*. New York: Basic Books.

Berch, B. 1980. Scientific management in the home: The empress's new clothes. *The Journal of American Culture* 3 (3): 440–45.

Bercuson, D. 1978. *Fools and wise men: The rise and fall of the One Big Union*. Toronto: McGraw-Hill Ryerson.

Bernard, B. 2004. Automaker allots $400m to retool Saturn Spring Hill. *Tennessean*, August 4.

Bernard, E. 1982. *The long distance feeling: A history of the telecommunications workers union*. Vancouver: New Star Books.

Bhagwati, J., A. Panagariya, and T. N. Srinivasan. 2004. The muddle over outsourcing. *Journal of Economic Perspectives* 18 (4): 93–114.

Bibby, A. 2002. A new union mould. *Financial Times*, February 27. http://global.factiva.com/ (accessed May 14, 2006).

——. 2004. The global mobility revolution. *UNI In Depth*, March 17. http://www.union-network.org/uniindep.nsf/51520e5cabae0089c12569b4004fe4c2/79fcc3c0a19684c9c1256e5a004517b6?OpenDocument (accessed June 4, 2006).

Bishop, T. 2007. Microsoft plans to open software center in B.C. *Seattle Post-Intelligencer*, July 5. http://seattlepi.nwsource.com/business/322652_msftvancouver06.html (accessed January 2, 2008).

Blok, A., and G. Downey, eds. 2003. *Uncovering labour in information revolutions*. Cambridge: Cambridge University Press.

Braverman, H. 1974. *Labor and monopoly capital*. New York: Monthly Review Press.

Brint, S. 2001. Professionals and the knowledge economy: Rethinking the theory of postindustrial society. *Current Sociology* 49 (4): 101–32.

Briskin, L., and P. McDermott. 1993. *Women challenging unions*. Toronto: University of Toronto Press.

Brodesser, C. 2004. Rings is SAG king. *Variety*, February 22. http://www.variety.com/ (accessed March 16, 2006).

Brody, D. 1993. *Workers in industrial America: Essays on the twentieth century struggle*, 2nd ed. New York: Oxford University Press.

Bronfenbrenner, K., R. L. Seeber, and R. A. Oswald, eds. 1998. *Organizing to win: New research on union strategies*. Ithaca, NY: ILR Press.

Brophy, E. 2006. System error: Labour precarity and collective organizing at Microsoft. *Canadian Journal of Communication* 31 (3): 619–38.

Broun, H. H., ed. 1941. *Collected edition of Heywood Broun*. New York: Harcourt.

Browning, L. 2007. Vermont becomes 'offshore' insurance haven. *New York Times*, April 4.

Buchanan, R., and S. Koch-Schulte. 2004. *Gender on the line: Technology, restructuring and the reorganization of work in the call centre industry*. Ottawa: Status of Women Canada. http://www.swc-fc.gc.ca/pubs/pubspr/0662281586/index_e.html (accessed June 14, 2007).

Buckley, C. 2004. Let a thousand ideas flower: China is a new hotbed of research. *New York Times*, September 13.

——. 2007. Britain's new super-union seeks to team up with U.S. counterpart. *Times Online*. April 6. http://business.timesonline.co.uk (accessed September 13, 2007).

Buhle, P. 2005. The Wobbly legend lives on in popular pulture. *The Chronicle of Higher Education* 51 (40): B11.

Burr, C. 1993. Defending 'the art preservative': Class and gender relations in the printing trades unions, 1850–1914. *Labour/Le Travail* 32 (Spring): 47–73.

Burris, B.H. 1998. Computerization of the workplace. *Annual Review of Sociology* 24: 141–57.

Byster, L. A., and T. Smith. 2006. From grassroots to global. In *Challenging the chip*, ed. T. Smith, D. A. Sonnenfeld, and D. N. Pellow, 111–19. Philadelphia: Temple University Press.

Canada Industrial Relations Board. 1999. *BCT.Telus et al.* Board File: 20051, Decision no. 36, October 28. Ottawa: Canada Industrial Relations Board.

——. 2000a. *BCT.Telus et al.* Board File: 20051-C, Decision no. 73, June 9. Ottawa: Canada Industrial Relations Board.

——. 2000b. *Island Telecom Inc. et al.* Board Files: 18403-C, 18729-C, Decision no. 59, February 24. Ottawa: Canada Industrial Relations Board.

——. 2000c. *TELUS Corporation*. Board File: 21298-C, Decision no. 94, October 19. Ottawa: Canada Industrial Relations Board.

——. 2003. *Canadian Broadcasting Corporation*. Board File: 20692-C, Decision no. 218, February 28. Ottawa: Canada Industrial Relations Board.

——. 2004a. *Telus Communications Inc. et al.* Board Files: 23742-C, 24063-C, 24160-C, Decision no. 271, April 8. Ottawa: Canada Industrial Relations Board.

——. 2004b. *Telus Communications Inc. et al.* Board File: 22076-C, Decision no. 278, June 24. Ottawa: Canada Industrial Relations Board.

——. 2005. *Société Radio-Canada*. Board File: 24325-C, Décision no. 308, February 16. Ottawa: Canada Industrial Relations Board.

Canadian Auto Workers. 2008. Who we are: Mergers. http://www.caw.ca/whoweare/mergers/index.asp (accessed February 2, 2008).

Canadian Film and Television Production Association. 2002. Canadian producers join forces to fight SAG's global rule. April 16. http://www.CFTPA.ca/ (accessed April 5, 2006).

Canadian Freelance Union. 2007. *CFU Newsletter*, Spring (1).

Canadian Media Guild. 2006. Last CBC-made youth show gets the axe. News release, May 17. http://www.cmg.ca/newsresults.asp?ID=827&SubjectID=37&BranchID=1 (accessed June 4, 2006).

Canwest. 2008 Corporate website. http://www.canwest.com/default.asp# (accessed January 15, 2008).

Cassedy, E., and K. Nussbaum. 1983. *9 to 5: The working woman's guide to office survival*. New York: Penguin Books.

Castells, M. 2000. *The rise of the network society*. Vol. 1 of *The information age: Economy, society and culture*. Oxford: Blackwell Publishers.

——. 2003. *The Internet galaxy*. London: Oxford University Press.

CAW. *See* Canadian Auto Workers.

CBC News. 2007. U.S. trade officials reject probe of Canadian film subsidies. October 22. http://www.cbc.ca/news/story/2007/10/22/trade-film.html (accessed October 22, 2007).

Center on Wisconsin Strategy. 2007. http://www.cows.org/ (accessed June 11, 2007).

Centre for Education and Communication, Communications Workers of America, Jobs with Justice, New Trade Union Initiative, and Young Professionals Collective. 2006. *Bi-national perspective on offshore outsourcing: A collaboration between Indian and US labour*. Washington, D.C.: Communications Workers of America.

CEP. *See* Communications, Energy and Paperworkers Union of Canada.

CFTPA. *See* Canadian Film and Television Production Association.

CFU. *See* Canadian Freelance Union.

Chaison, G. 1986. *When unions merge*. Lexington, MA: Lexington Books.

——. 2001. Union mergers and union revival: Are we asking too much or too little? In *Rekindling the movement: Labor's quest for relevance in the twenty-first century*, eds. L. Turner, H. C. Katz, and R. W. Hurd, 238–55. Ithaca, NY: Cornell University Press.

Chakravartty, P. 2005. Weak winners of globalization: Indian H-1B workers in the American information economy. *AAPI Nexus* 3 (2): 59–84. http://www.aasc.ucla.edu/aascpress/nexuscollection.htm (accessed May 2, 2008).

Chepesiuk, R. 2005. Outsourcing the Western media. *The Global Journalist*, First Quarter.

Cieply, M. 2007. Hollywood officials seek end to system of paying residuals. *New York Times*, July 12.

Cieply, M., and B. Barnes. 2008. After the writers' strike. *New York Times*, February 11.

CIRB. *See* Canada Industrial Relations Board.

Cirque du Soleil. 2007. Cirque du Soleil at a Glance. http://www.cirquedusoleil.com/CirqueDuSoleil/en/Pressroom/cirquedusoleil/factsheets/default.htm (accessed December 21, 2007).

Clawson, D. 2003. *The next upsurge: Labor and the new social movements*. Ithaca, NY: ILR Press.

Cleaver, H. 1993. An Interview with Harry Cleaver. Original interview by Massimo De Angelis in Italian. *vis-a-vis* 1 (Autumn). Trans. and republished online at http://www.geocities.com/CapitolHill/3843/cleaver.html (accessed June 17, 2007).

CMG. *See* Canadian Media Guild.

Cockburn, C. 1983. *Brothers: Male dominance and technological change*. London: Pluto Press.

Coles, A. 2006. Acting in the name of culture? Organized labour campaigns for Canadian dramatic programming. *Canadian Journal of Communication* 30 (3): 519–40.

Collins, G. 2005. The gendered nature of mergers. *Gender, Work & Organization* 12 (3): 270–90.

Communications, Energy and Paperworkers Union of Canada. 2007. *Canadian media study: Voices from the newsroom*. Ottawa: Communications, Energy and Paperworkers Union of Canada.

Communications Workers of America. 2007. *Speed matters: Affordable high-speed Internet for all*. Washington, D.C.: Communications Workers of America.

Conell, C., and K. Voss. 1990. Formal organization and the fate of social movements: Craft association and class alliance in the Knights of Labor. *American Sociological Review* 55 (2): 255–69.

Connell, A. 2003. AFTRA/SAG failed consolidation and its impact on New England's acting community. *New England Entertainment Digest*, July 29. http://www.jacneed.com/ (accessed March 16, 2006).

Cooper, M. 2000. Runaway shops. *Nation*, March 16.

———. 2001. Residual anger. *Nation*, April 2. http://www.thenation.com/ (accessed August 31, 2007).

Courtney, M., President, Washington Alliance of Technology Workers. 2005. Interview with Enda Brophy. Seattle, July 6.

Courtney, M. 2007a. Tech workers score victory with House pass TAA. *WashTech News*, October 31. http://www.washtech.org/news/industry/display.php?ID_Content=5214 (accessed January 4, 2008).

———. 2007b. AT&T Mobility outsources tech jobs to IBM: Jobs could end up in India. *WashTech News*, October 16.

Cowan, R. S. 1979. The industrial revolution in the home: Household technology and social change in the 20th century. In *Dynamos and virgins revisited: Women and technological change in history*, ed. M. M. Trescott, 205–32. Metuchen, NJ: Scarecrow Press.

———. 1983. *More work for mother: The ironies of household technology from the open hearth to the microwave*. New York: Basic Books.

COWS. *See* Center on Wisconsin Strategy.

Cranford, C., L. Vosko, and N. Zukewich. 2003. Precarious employment in the Canadian labour market: A statistical portrait. *Just Labour: A Canadian Journal of Work and Society* Fall: 6–22.

Craypo, C. 1997. The impact of changing corporate strategies on communities, unions, and workers in the United States of America. *Journal of Law and Society* 24 (1): 10–25.

Craypo, C., and B. Nissen, eds. 1993. *Grand designs: The impact of corporate strategies on workers, unions and communities*. Ithaca, NY: ILR Press.

CWA. *See* Communications Workers of America.

Dalla Costa, M., and G. Dalla Costa, eds. 1999. *Women, development, and labor of reproduction: Struggles and movements*. Trenton, NJ: Africa World Press.

Dark, T. E. 1999. Debating decline: The 1995 race for the AFL-CIO presidency. *Labor History*, 40 (3): 323–43.

Davenport, T. H. 2005. *Thinking for a living: How to get better performance and results from knowledge workers*. Boston: Harvard Business School Press.

Davis, M. 2001. *Late Victorian holocausts: El Nino famines and the making of the Third World*. London: Verso.

Dawes, C. J. 1987. *The relative decline of international unionism in Canada since 1970*. School of Industrial Relations Research Essay Series, no. 19. Kingston, ON: Queen's University.

Day, T. 1992. Capital-labor substitution in the home. *Technology and Society* 33 (2): 302–27.

Day, W. W. 2007. Commodification of creativity: Reskilling computer animation labor in Taiwan. In *Knowledge workers in the information society*, eds. C. McKercher and V. Mosco, 85–99. Lanham, MD: Lexington Books.

De Groot, G., and M. Schrover, eds. 1995. *Women workers and technological change in Europe in the nineteenth and twentieth centures*. London: Taylor & Francis.

Deibel, M. 1999. Temporary workers want more fruit of labor. *The Patriot Ledger*, July 8.

Delacourt, S. 2005. CBC lockout shakes Liberals. *Toronto Star*, September 25.

Denning, M. 1996. *The cultural front*. New York: Verso.

Deshpande, J. V. 2006. Towards a knowledge-based society. *Economic and Political Weekly* (September 16): 3933–36.

Deuze, M. 2007. *Media work*. Cambridge, UK: Polity.

DGA. *See* Directors Guild of America.

Dhillon, A. 2006. Indian call centres + American students = offshore maths tutoring. *The Age*, March 20. http://www.theage.com.au/news/business/indian-call-centres—american-students—offshore-maths-tutoring/2006/03/19/1142703218230.html (accessed September 3, 2007).

Directors Guild of America. 2004. DGA applauds the signing of the American Jobs Creation Act into law. October 11. http://www.dga.org/index2.php3?chg= (accessed March 20, 2007).

Dowling, E., R. Nunes, and B. Trott. 2007. Immaterial and affective labor: Explored. *Ephemera: Theory and Politics in Organization* 7 (1): 1–7. http://www.ephemeraweb.org/journal/7-1/7-1index.htm (accessed January 24, 2008).

Drucker, P. 1959. *Landmarks of tomorrow*. New York: Harper.

———. 1993. *Post-capitalist society*. New York: Harper Business.

———. 2001. The new workforce. *Economist*, November 3.

Duffy, P. 2000. *The skilled compositor, 1850–1914: An aristocrat among working men*. Aldershot, UK: Ashgate Publishing.

Dunn, M. L., Jr., Assistant National Executive Director of the American Federation of Television and Radio Artists. 2006. Interview with authors. Los Angeles, February 6.

Duxbury, L., and C. Higgins. 2001. *Work-life balance in the new millennium: Where are we? Where do we need to go?* Ottawa: Canadian Policy Research Networks.

Dyer-Witheford, N. 1999. *Cyber-Marx: Cycles and circuits of struggle in high technology capitalism*. Urbana and Chicago: University of Illinois Press.

Economic Policy Institute. 2004. *Offshoring*. Washington, D.C.: Economic Policy Institute.

———. 2006. *Offshoring*. Washington, D.C.: Economic Policy Institute. Updated May 2006. http://www.epinet.org/content.cfm/issueguide_offshoring (accessed February 5, 2007).

Economist. 2005a. Into Africa. August 27.

———. 2005b. Time to bring it back home? March 3.

———. 2006. Freelancers of the world, unite! November 9.

Editor and Publisher. 2008. Pressmart to provide digital editions for Torstar's Metroland West. January 22. http://www.editorandpublisher.com/eandp/index.jsp (accessed January 23, 2008).

Edwards, L. N., and E. Field-Hendrey. 2002. Home-based work and women's labor force decisions. *Journal of Labor Economics* 20 (1): 170–200.

Eggerton, J. 2005. Unions consolidate to face merged media. *Broadcasting & Cable*, October 11.

Eisenstein, Z. 1998. *Global obscenities: Patriarchy, capitalism, and the lure of cyberfantasy*. New York: New York University Press.

———. 2004. *Against empire: Feminisms, racism, and the West*. London: Pluto Press.

Elber, L. 2008. Writers vote to end 3-month walkout. *washingtonpost.com*, February 13. http://www.washingtonpost.com/wp-dyn/content/article/2008/02/12/AR2008021200546.html?hpid=moreheadlines (accessed February 18, 2008).

Elmer, G., and M. Gasher, eds. 2005a. *Contracting out Hollywood: Runaway productions and foreign location shooting*. Lanham, MD: Rowman & Littlefield.

———. 2005b. Introduction: Catching up to runaway productions. In *Contracting out Hollywood*, eds. G. Elmer and M. Gasher, 1–18. Lanham, MD: Rowman & Littlefield.

Engardio, P. 2006. The future of outsourcing. *Business Week*, January 30.

Ephemera: Theory and Politics in Organization. 2007. Vol. 7, no. 1 (February). http://www.ephemeraweb.org/journal/7-1/7-1index.htm (accessed January 24, 2008).

EPI. *See* Economic Policy Institute.

Estevenin, K., Organizer, Washington Alliance of Technology Workers. 2005. Interview with Enda Brophy. Seattle, July 8.

Fair, P., Deputy National Executive Director of the Screen Actors Guild. 2006. Interview with authors. Los Angeles, February 3.

Fan, M. 2000. California labor leaders turn focus toward immigrant workers. *Knight Ridder Tribune Business News*, January 21.

Felstead, A., N. Jewson, and S. Walters. 2003. Managerial control of employees working at home. *British Journal of Industrial Relations* 41 (2): 241–64.

Ferner, A., and M. Terry. 1997. United Kingdom. In *Telecommunications: Restructuring work and employment relations worldwide*, ed. H. C. Katz, 89–121. Ithaca, NY: Cornell University Press.

Financial Express. 2004. Forum to tackle unfair HR policies in BPO industry. July 28. http://global.factiva.com/ (accessed May 14, 2006).

———. 2007. Hungry tiger, dancing elephant. Reprinted from *Economist*, April 9. http://www.financialexpress.com/news/story/163446/ (accessed September 5, 2007).

Fiorito, J., P. Jarley, and J. T. Delaney. 1995. National union effectiveness in organizing: Measures and influences. *Industrial and Labor Relations Review* 48 (4): 613–35.

Fletcher, B. J., and R. W. Hurd. 1998. Beyond the organizing model: The transformation process in local unions. In *Organizing to win: New research on union strategies*, eds. K. Bronfenbrenner, R. L. Seeber, and R. A. Oswald, 37–53. Ithaca, NY: ILR Press.

Florida, R. 2002. *The rise of the creative class*. New York: Basic Books.

Fones-Wolf, E., and C. Fones-Wolf. 2007. Labor off the air: The Hearst Corporation, cross ownership and the union struggle for media access in San Francisco. In *Knowledge workers in the information society*, eds. C. McKercher and V. Mosco, 1–18. Lanham, MD: Lexington Books.

Fones-Wolf, E., and K. Fones-Wolf. 1983. The war at Mingo Junction: The autonomous workman and the decline of the Knights of Labour. *Ohio History* 92: 36–51.

Fong, P. 2004. India's largest IT service provider expands west. *Vancouver Sun*, October 14.

Foote, N., and P. K. Hatt. 1953. Social mobility and economic advancement. *American Economic Review* 43 (2): 364–79.

Foster, J. B. 1988. The fetish of Fordism. *Monthly Review* 39 (10): 14–33.

Fourastié, J. 1954. Predicting economic changes in our time. *Diogenes* 2 (5): 14–38.

Fox, N. 2002. *Against the machine: The hidden Luddite tradition in literature, art and individual lives.* Washington, D.C.: Island Press.

Fraser, J. A. 2001. *White-collar sweatshop: The deterioration of work and its rewards in corporate America*. New York: W.W. Norton & Company.

Frauenheim, E. 2005. Will India's outsourcers rule China's roost? *CNET News*, February 4. http://news.com.com/2061-1022-5564260.html (accessed January 2, 2008).

Frederick, C. 1913. *The new housekeeping: Efficiency studies in home management.* Garden City, NY: Doubleday, Page & Company.

———. 1919. *Household engineering: Scientific management in the home*. Chicago: American School of Home Economics.

Freeze, C. 2006. Bell settles pay equity suit. *Globe and Mail*, May 16.

Friedman, T. L. 2005. *The world is flat: A brief history of the twenty-first century*. New York: Farrar, Strauss and Giroux.

———. 2006. Latin America's choice. *New York Times*, June 21.

Fryer, A. 1998. A budding union for temps. *Seattle Times*, April 30.

Galt, V. 2005. India's ICICI catches the onshore wave. *Globe and Mail,* March 28, Report on Business.

Gapasin, F., and H. Wial. 1998. The role of central labour councils for union organizing in the 1990s. In *Organizing to win: New research on union strategies*, eds. K. Bronfenbrenner, R. L. Seeber, and R. A. Oswald, 54–67. Ithaca, NY: ILR Press.

Gates, B. 1996. *The road ahead*. With N. Myhrvold and P. Rinearson. New York: Penguin.

———. 1999. *Business @ the speed of thought*. New York: Warner Books.

GE. 2008 Corporate website. http://www.ge.com/company/businesses/factsheets/corporate.html (accessed January 15, 2008).

Gerstein, I. 1979. Domestic work and capitalism. *Radical America* 7 (4): 101–28.

Gibson, R. 2007. Letter to the editor. *New York Times*, December 31.

Gilbreth, L. M. 1927. *The home-maker and her job*. New York: Appleton.

Giridharadas, A. 2007a. Lobbying in U.S., Indian firms present an American face. *New York Times*, September 4. http://www.nytimes.com/2007/09/04/business/world business/04outsource.html (accessed January 4, 2008).

——. 2007b. Outsourcing works, so India is exporting jobs. *New York Times*, September 25. http://www.nytimes.com/2007/09/25/business/worldbusiness/25outsource.html (accessed September 26, 2007).

Goldstein, P. 2005. His union divided cannot stand. *Los Angeles Times*, October 4.

Gouldner, A. 1979. *The future of intellectuals and the rise of the new class*. New York: Seabury Press.

Granville, S. 1999. Women workers and the new economy. Paper presented to the conference "American Labor and the New Economy—A Day of Dialogue," sponsored by Social Democrats, USA, and the League for Industrial Democracy, January 22, Washington, D.C. http://www.newecon.org/womenwork_sg.html (accessed October 13, 2005).

Greene, A., and G. Kirton. 2003. Possibilities for remote participation in trade unions: Mobilizing women activists. *Industrial Relations Journal* 34 (4): 319–33.

Greenhouse, S. 2008. Union membership sees biggest rise since '83. *New York Times*, January 26. http://www.nytimes.com/2008/01/26/us/26labor.html?scp=1&sq=greenhouse&st=nyt (accessed January 26, 2008).

Grey, D. 1990. Report to the printing, publishing and media workers sector. Presented at the annual convention of the Communications Workers of America, June 11–13, Toronto.

Guider, E. 2008. Hollywood strike cost $2.5 billion: Report. *Hollywood Reporter*, February 20. Reprinted at www.washingtonpost.com (accessed February 25, 2008).

Gupta, T. D. 2006. Racism/anti-racism, precarious employment, and unions. In *Precarious employment: Understanding labor market insecurity in Canada*, ed. L. Vosko, 318–34. Montreal: McGill-Queen's University Press.

Haines, P., and G. Klein. 1982. Citizens and unions respond. In *Community and capital in conflict: Plant closing and job loss*, eds. J. C. Raines, L. Berson, and D. Grace, 234–54. Philadelphia: Temple University Press.

Halford, S. 2005. Hybrid workspace: Re-spatialisations of work, organisation and management. *New Technology, Work and Employment* 20 (1): 19–33.

Hall, S., C. Critcher, T. Jefferson, J. Clarke, and B. Roberts, 1978. *Policing the crisis: Mugging, the state and law and order*. London: Macmillan.

Hanc, J. 2007. Your computers are down? No problema. *New York Times*, November 14.

Haraway, D. 1991. A cyborg manifesto: Science, technology, and socialist-feminism in the late twentieth century. In *Simians, cyborgs and women: The reinvention of nature*, D. Haraway, 149–81. New York: Routledge.

Hardt, M., and A. Negri. 2000. *Empire*. Cambridge, MA: Harvard University Press.

——. 2004. *Multitude: War and democracy in the age of empire*. New York: Penguin Press.

Harman, C. 2003. Initial statement: Chris Harman. In the Harman-Hardt debate at the World Social Forum, January 25, in Porto Alegre, Brazil. http://www.forumsocialmundial.org.br/dinamic.php?pagina=of_globalise_ing (accessed February 23, 2008).

Harvey, D. 2006. *Spaces of global capitalism: Towards a theory of uneven geographical development*. New York: Verso.

Head, S. 2003. *The new ruthless economy: Work and power in the digital age*. Oxford: Century Foundation/Oxford University Press.

Heckscher, C., and D. Palmer. 1993. Associational movements and employment rights: An emerging paradigm. *Research in the Sociology of Organizations* 12: 279–309.

Heher, A. 2007. Bloggers consider forming labour union. *Globe and Mail*, August 7. http://www.theglobeandmail.com/servlet/story/RTGAM.20070807.wgtunion0807/EmailBNStory/Technology/ (accessed August 8, 2007).

Herman, E. S., and N. Chomsky. 2002. *Manufacturing consent: The political economy of the mass media*. New York: Pantheon.

Hira, R. 2007. *Outsourcing America's technology and knowledge jobs*. Economic Policy Institute, briefing paper no. 187. Washington, D.C.: Economic Policy Institute.

Hobsbawm, E. J. 1964. *Labouring men: Studies in the history of labour*. London: Weidenfeld & Nicolson.

Holstein, W. J. 2005a. Finding a sea less crowded. *New York Times*, January 30. http://www.nytimes.com/2005/01/30/business/yourmoney/30advi.html?pagewanted=print&position (accessed January 1, 2008).

———. 2005b. Cutting the losses from outsourcing. *New York Times*, July 3.

Horn, J. 2004. A 'Cold' war over foreign filming; Frustrated by job losses, industry workers launch an Internet campaign against 'Cold Mountain' for choosing Romania as its American stand-in. *Los Angeles Times*, February 4.

Horwitch, L. 2005. AFTRA tackles finances. *Backstage.com*, July 22. http://www.backstage.com/ (accessed February 10, 2008).

Hossfeld, K. J. 1990. "Their logic against them": contradictions of sex, race, and class in Silicon Valley. In *Women workers and global restructuring*, ed. K. Ward, 149–78. Ithaca, NY: ILR Press.

Howell, N. 2006. Global organizing for call centre workers. In *ILO, Labour Education 2006: Globalization and the Sectors*, ed. F. van Leeuwen. http://www.ilo.org/public/english/dialogue/actrav/publ/142/142.pdf (accessed June 5, 2007).

Huws, U. 2003. *The making of a cybertariat: Virtual work in a real world*. New York: Monthly Review Press.

———. 2006. What will we do? The destruction of occupational identities in the 'knowledge-based economy.' *Monthly Review* 57 (8). http://www.monthlyreview.org/0106huws.htm (accessed July 9, 2007).

ICTUR. *See* International Centre for Trade Unions Rights.

ILO. *See* International Labour Organization.

Indiatimes. 2007. Infocrossing bags $407-m contract. December 6. http://www.economictimes.indiatimes.com/Infotech/ITeS/Infocrossing_bags_407-m_contract/articleshow/2599185.cms (accessed December 23, 2007).

International Centre for Trade Unions Rights. 2005. *Trade Unions of the World*, 6th ed. London: John Harper Publishing.

International Labour Organization. 2001. *Promoting gender equality—a resource kit for trade unions.* Geneva: Gender Promotion Programme, International Labour Office, ILO.http://www.ilo.org/public/english/employment/gems/eeo/tu/cha_1.htm#1.%20Trade%20unions%20and%20gender%20equality (accessed September 27, 2007).

———. 2002. Tripartite meeting on employment, employability and equal opportunities in the postal and telecommunications services. ILO Sectoral Activities Programme. http://www.ilo.org/public/english/dialogue/sector/techmeet/tmpts02/tmpts-r.pdf (accessed July 1 2006).

Jackson, A. 2003. Is work working for women? Canadian Labour Congress, research paper no. 22. Ottawa: Canadian Labour Congress.

Jacobson, S. 1960. The fourth estate: A study of the American Newspaper Guild. PhD diss., New School for Social Research.

Jennings, A. 2007. Thomas the Tank engine toys recalled because of lead paint. *New York Times*, June 15.

Jennings, P. J. 2006. Philip J. Jennings' address to LBC-NVK congress. *UNI News Flash*, March 24. http://www.union-network.org/uniflashes.nsf/40d8481f779124d4c12568db00372009/744d1cf60ece0b28c125713f00472e50?OpenDocument (accessed July 1, 2006).

Johnson, B. 2008. Media work force sinks to 15-year low: Newspaper slump and the shift to digital, direct take toll on employment. *Advertising Age,* February 18. www.adage.com (accessed February 25, 2008).

Johnston, P. 2001. Organize for what? In *Rekindling the movement: Labor's quest for relevance in the 21st century*, ed. L. Turner, H. C. Katz, and R. W. Hurd, 27–58. Ithaca, NY: Cornell University Press.

Kapur, J. 2007. "New" economy/old labor: Creativity, flatness, and other neo-liberal myths. In *Knowledge workers in the information society*, ed. C. McKercher and V. Mosco, 163–75. Lanham, MD: Lexington Books.

Katz, H. C. 1993. The decentralization of collective bargaining: A literature review and comparative analysis. *Industrial and Labor Relations Review* 47 (1): 3–22.

———. 1997. Introduction and comparative overview. In *Telecommunications: Restructuring work and employment relations worldwide*, ed. H. C. Katz, 2–28. Ithaca, NY: Cornell University Press.

Kaufman, J. 2001. Rise and fall of a nation of joiners: The Knights of Labor revisited. *Journal of Interdisciplinary History* 31 (4): 553–79.

Kealey, G. S. 1986. Work control, the labour process and nineteenth-century Canadian printers. In *On the job: Confronting the labour process in Canada*, ed. C. Heron and R. Story, 75–101. Kingston, ON: McGill-Queen's University Press.

Kealey, G. S., and B. D. Palmer. 1987. *Dreaming of what might be: The Knights of Labor in Ontario 1880–1900*. Toronto: New Hogtown Press.

Keefe, J. H., and R. Batt. 1997. United States. In *Telecommunications: Restructuring work and employment relations worldwide*, ed. H. C. Katz, 31–88. Ithaca, NY: Cornell University Press.

Keefe, J., and K. Boroff. 1994. Telecommunications labor-management relations: One decade after the AT&T divestiture. In *Contemporary collective bargaining in the private sector*, ed. P. B. Voos, 303–72. Madison, WI: Industrial Relations Research Association.

Kelber, H., and C. Schlesinger. 1967. *Union printers and controlled automation*. New York: Free Press.

Kellerman, S. 2007. A century of writing on the IWW: An annotated bibliography of books on the Industrial Workers of the World. http://www.iww.org/en/culture/library (accessed August 27, 2007).

Kiefer, P. 2003. SAG-AFTRA merger proposal shot down. *Hollywood Reporter*, July 2. http://www.hollywoodreporter.com/hr/search/article_display.jsp?vnu_content_id=1925686 (accessed February 23, 2008).

Kimbrough, M., National Recording Secretary of the American Federation of Television and Radio Artists. 2006. Interview with authors. Los Angeles, February 6.

Kimeldorf, H. 1999. *Battling for American labor: Wobblies, craft workers, and the making of the union movement*. Berkeley and Los Angeles: University of California Press.

Kiss, S., and V. Mosco. 2005. Negotiating electronic surveillance in the workplace: A study of collective agreements in Canada. *Canadian Journal of Communication* 30 (4): 549–64.

Kleinegger, C. 1987. Out of the barns and into the kitchens. In *Women, work and technology*, ed. B. D. Wright, 162–78. Ann Arbor: The University of Michigan Press.

Kline, S., N. Dyer-Witheford, and G. de Peuter. 2003. *Digital play*. Montreal: McGill-Queen's University Press.

Klinenberg, E. 2007. *Fighting for air: The battle to control America's media*. New York: Metropolitan Books.

Knights of Labor. 1878. *Preamble to the constitution of the Knights of Labor*. http://www.historytools.org/sources/knights.html (accessed August 24, 2007).

Koch, J. 2005. Thinking offshore through. *Executive Action*, February 15.

Kohl, G., Research Director for the Communications Workers of America. 2007. Interview with authors, Washington, D.C., March 22.

Korvajarvi, P. 2003. Women and technological pleasure at work. In *Information society and the workplace*, ed. T. Heiskanen and J. Hearn, 125–42. London: Routledge.

Kramer, L. 2007. Training local workers instead of outsourcing. *New York Times*, August 5.

Krishnan, H. A., and D. Park. 2002. The impact of workforce reduction on subsequent performance in major mergers and acquisitions: An exploratory study. *Journal of Business Ethics* 55 (4): 285–92.

Krupat, K. 1999. Out of labor's Dark Age: Sexual politics comes to the workplace. *Social Text* 17 (4): 9–29.

Kumar, D. 2007a. Globalization and workers' power: The struggle for hegemony during the 1997 UPS Strike. In *Knowledge workers in the information society*, eds. C. McKercher and V. Mosco, 267–84. Lanham, MD: Lexington Books.

———. 2007b. *Outside the box: Corporate media, globalization and the UPS strike*. Urbana: University of Illinois Press.

Kumar, N., and N. Verghese. 2004. Money for nothing and calls for free. *Global Policy Forum*, February 17.

Langberg, M. 2005. Union leader makes tech pitch. *New York Times*, March 23.

Lareau, L. 2001. CBC/SRC: The latest on the CIRB hearings re: "one big union." Statement from the president of the Canadian Media Guild, November 2. http://

www.cmg.ca/newsresults.asp?ID=51&BranchID=1&SubjectID=24 (accessed July 2, 2007).

Laut, A. C. 1912/2003. *Am I my brother's keeper? A study of British Columbia's labor and Oriental problems*. Collected articles first published in *Saturday Night*, reprinted with an introduction by M. Leier. Vancouver: Subway Books.

Lawrence, E. 1994. *Gender and trade unions*. London: Taylor & Francis.

Lazzarato, M. 1997. *Lavoro immateriale: Forme di Vita e produzione di soggettività*. Verona: Ombre Corte.

LeFevre, W. 2005. In the shadow of the IWW. http://www.reuther.wayne.edu/exhibits/iww.html (accessed September 13, 2005).

Leier, M. 1989. *Where the Fraser River flows: The Industrial Workers of the World in British Columbia*. Vancouver: New Star Books.

Leonhardt, D. 2007a. Economic scene: Lessons Thomas could learn. *New York Times*, October 24.

———. 2007b. He's happier, she's less so. *New York Times*, September 26.

Levi, M., D. J. Olson, and E. Steinman. 2002–2003. Living-wage campaigns and laws. *WorkingUSA* 6 (3): 111–32.

Liker, J. 2004. *The Toyota way*. New York: McGraw-Hill.

Lin, J. 2006. The teaching profession: Trends from 1999 to 2005. *Education Matters* 3(4).

Lipset, S. M., M. A. Trow, and J. S Coleman. 1956. *Union democracy: The internal politics of the International Typographical Union*. Glencoe, IL: Free Press.

Liu, A. 2004. *The laws of cool: Knowledge work and the culture of information*. Chicago: University of Chicago Press.

Livingstone, D. W., and P. H. Sawchuk. 2004. *Hidden knowledge: Organized labor in the information age*. Aurora, ON: Garamond Press.

Lohr, S. 2004. A dissenter on outsourcing states his case. *International Herald Tribune*, September 7.

———. 2007a. IBM showing that giants can be nimble. *New York Times*, July 18.

———. 2007b. Hello, India? I need help with my math. *New York Times*, October 31.

Lomas, N. 2007. Survey finds increasing uncertainty over offshoring. *CNET News .com,* August 22. http://www.news.com/ (accessed September 10, 2007).

London Free Press. 1966. *Communications in the community*. Limited souvenir edition to commemorate opening of a new building. London: London Free Press.

Longstaff, P. 2002. *The communications toolkit*. Cambridge, MA: MIT Press.

Louie, M. C. Y. 2001. *Sweatshop warriors: Immigrant women workers take on the global economy*. Cambridge, MA: South End Press.

Lovell, V., H. Hartmann, and J. Koski. 2006. *Making the right call: Jobs and diversity in the communications and media sector*. Washington, D.C.: Institute for Women's Policy Research.

Machlup, F. 1962. *The production and distribution of knowledge in the United States*. Princeton, NJ: Princeton University Press.

Magder, T., and J. Burston 2001. Whose Hollywood? Changing forms and relations inside the North American entertainment industry. In *Continental order? Integrating North America for cybercapitalism*, eds. V. Mosco and D. Schiller, 207–34. Lanham, MD: Rowman & Littlefield.

Mahapatra, R. 2007. 'Lifestyle' diseases hit India's high tech workers. *The Guardian*, September 14.

Mann, C. 2003. *Globalization of IT services and white collar jobs: The next wave of productivity growth*. Institute for International Economics, policy brief 03-11. Washington, D.C.: Institute for International Economics.

Marquez, J. 2006. Unions' global end run. *Workforce Management* 85 (2). http://global.factiva.com/ (accessed May 15, 2006).

Marschall, D. 2002. Internet technologists as an occupational community: Ethnographic evidence. *Information, Communication & Society* 5 (1): 51–69.

Marshall, T. H. 1964. *Class, citizenship, and social development: Essays*. Westport, CT: Greenwood Press.

Martin, C. R. 2004. *Framed! Labor and the corporate media*. Ithaca, NY: ILR Press.

——. 2007. "Writing off workers: The decline of the U.S. and Canadian labor beats," in *Knowledge workers in the information society*, eds. C. McKercher and V. Mosco, 19–35. Lanham, MD: Lexington Books.

Marx, K. 1973. *Grundrisse*. Middlesex, UK: Penguin Books.

Marx, L. 1964. *The machine in the garden: Technology and the pastoral ideal in America*. New York: Oxford University Press.

Mattingly, M. J., and S. M. Bianchi. 2003. Gender differences in the quantity and quality of free time: The U.S. experience. *Social Forces* 81 (3): 999–1030.

McCarthy, J. C. 2002. *3.3 million service jobs to go offshore*. Cambridge, MA: Forrester Research.

McDougall, P. 2005. Exclusive: Gartner predicts huge increase in offshore outsourcing by 2015. *Information Week*, March 31.

McKercher, C. 2002. *Newsworkers unite: Labor, convergence and North American newspapers*. Lanham, MD: Rowman & Littlefield.

McKercher, C., and V. Mosco, eds. 2006. The Labouring of Communication. Special issue of the *Canadian Journal of Communication*, Vol. 31, No 3.

——, eds. 2007. *Knowledge workers in the information society*. Lanham, MD: Lexington Books.

McLaughlin, L., and H. Johnson. 2007. Women and knowledge work in the Asia Pacific: Complicating technological empowerment. In *Knowledge workers in the information society*, eds. C. McKercher and V. Mosco, 249–66. Lanham, MD: Lexington Books.

McNary, D. 2005. SAG looks to law on runaway prod'n. *Variety*, October 26. http://www.ftac.net/html/sag-var-10-26-05.html (accessed March 20, 2007).

——. 2007a. SAG condemns guild quitters: Board cracks down on 'fi-core.' *Variety*, May 21. http://www.variety.com/article/VR1117965548.html?categoryid=1066&cs=1 (last accessed, July 9, 2007).

——. 2007b. SAG and AFTRA trade accusations: Contract poaching a hot topic. *Variety.com*, October 17.

——. 2007c. SAG, AFTRA prepare for battle: Unions threaten to fly solo during negotiations. *Variety.com*, October 22.

——. 2007d. Unity push for SAG, AFTRA: Campaign aims to revive merger of unions. *Variety.com*, October 9.

McWilliams, C. 1939. *Factories in the field*. Boston: Little, Brown.

Meehan, E. R. 2002. Gendering the commodity audience: Critical media research, feminism, and political economy. In *Sex and money: Feminism and political economy in the media*, eds. E.R. Meehan and E. Riordan, 209–22. Minneapolis: University of Minnesota Press.

Meehan, E. R., and E. Riordan, eds. 2002. *Sex and money: Feminism and political economy in the media*. Minneapolis: University of Minnesota Press.

Melnicer, S. 2005. The CBC lockout. Am I an 'us' or a 'them'? *The Globe and Mail*, August 17. http://www.theglobeandmail.com/servlet/story/RTGAM.20050817.wcomment0818/PPVStory?URL_Article_ID=RTGAM.20050817.wcomment0818&DENIED=1 (accessed November 18. 2007).

Menzies, H. 1996. *Whose brave new world?* Toronto: Between the Lines.

Miller, A. L. 2006. AFTRA, the monkey on SAG actors' backs. *SAGWatchdog.com*, January 18. http://sagwatchdog.com/cgi-bin/index.cgi/read/576 (accessed February 10, 2008).

Miller, J. M., ed. 1998. *CWA at fifty: A pictorial history of the Communications Workers of America*. Washington, D.C.: Communications Workers of America.

Mishra, B. R. 2007. Meet the man behind the ITeS union. *Rediff India Abroad*, February 1. http://www.rediff.com/money/2007/feb/01ites.htm (accessed June 4, 2006).

Moberg, D. 2000. Temp slave revolt: Contingent workers of the world unite. *In These Times*, July 10.

Monbiot, G. 2003. The flight to India. *Guardian*, October 21.

Monitor Company. 1999. U.S. runaway film and television production study. Commissioned by the Screen Actors Guild and Directors Guild of America. http://www.dga.org/news/pr_runaway.pdf (accessed March 20, 2007).

Moore, K., and M. Hertz. 2008. The world's not so flat. *Globe and Mail*, January 2. http://www.theglobeandmail.com/servlet/story/RTGAM.20080102.wwebexc0103/EmailBNStory/Front/home (accessed January 4, 2008).

Moore, T. 2005. Big labour fights for survival. *Winnipeg Free Press*, September 4. http://global.factiva.com/ (accessed July 12, 2006).

Morgan, O. 2000. The unions: Open shop in the new economy. *Observer*, September 4. http://global.factiva.com/ (accessed July 11, 2006).

———. 2006. Amicus, IG-Metall and two US labor groups join forces to confront the power of the multinationals. *Observer*, December 31.

Morgan, R., President of the Los Angeles local of the American Federation of Television and Radio Artists. 2006. Interview with authors. Los Angeles, February 6.

Mosco, V. 1994. Getting it right with computer communication: A case study of the United Services Automobile Association. Program on Information Resources Policy, Harvard University, December, P-94-9.

———. 1996. *The political economy of communication*. London: Sage Publications.

———. 1999. Report on BCT.Telus (a merged corporation formed from BC Telecom and Telus Corporation) and the Telecommunications Workers Unions (TWU) re: Application pursuant to sections 18, 18.1, 35, 44, 45 and 46 of the *Canada Labour Relations Code*. November.

———. 2002. Report on Canada Industrial Relations Board File No.22076-C re: Application pursuant to sections 18, 18.1, 35, 44, 45 & 46 of the *Canada Labour Re-*

lations Code between the Telecommunications Workers Union and Telus Communications Inc. Submitted to the Canada Industrial Relations Board, June.

———. 2004. *The digital sublime*. Cambridge, MA: MIT Press.

———. 2005. Here today, outsourced tomorrow: Knowledge workers in the global economy. *Javnost/The Public* 12 (2): 39–56.

Mosco, V., and C. McKercher. 2006. Convergence bites back: Labour struggles in the Canadian communication industry. *Canadian Journal of Communication* 31 (3): 733–52.

Mosco V., and J. Wasko, eds. 1983. *Labor, the working class and the media*. Norwood, NJ: Ablex Publishing.

Mosco V., E. Zureik, and C. Lochhead. 1989. Telephone workers' reaction to the new technology. *Industrial Relations* 44 (3): 507–29.

Munro, A. 2001. A feminist trade union agenda? The continued significance of class, gender and race. *Gender, Work & Organization* 8 (4): 454–71.

Murray, G. 1995. Unions: Membership, structures and action. In *Union-management relations in Canada*, 3rd ed., ed. M. Gunderson and A. Ponak, 159–94. Don Mills, ON: Addison-Wesley.

Nagy, I. 2007. Labor strife and carnival symbolism. In *Knowledge workers in the information society*, ed. C. McKercher and V. Mosco, 285–98. Lanham, MD: Lexington Books.

Nandgaonkar, S. 2005. Workers of the world unite. *Telegraph* (Calcutta), January 20.

National Writers Union. 2007. http://www.nwu.org/ (accessed June 1, 2007).

Negroponte, N. 1996. *Being digital*. Cambridge, MA: MIT Press.

New India Express. 2005. Radical hypocrisy: It's time the Left ended this agony. September 30. http://www.newindpress.com/newspages.asp?page=f&Title=First+Editorial&aDate=9%2F30%2F2005 (accessed September 12, 2007).

Nichols, J., and R. W. McChesney. 2005. *Tragedy and farce: How the American media sell wars, spin elections, and destroy democracy*. New York: New Press.

Nissen, B. 2000. Living wage campaigns from a 'social movement' perspective: The Miami case. *Labor Studies Journal* 25 (3): 29–50.

———, ed. 2002. *Unions in a globalized environment: Changing borders, organizational boundaries and social roles*. Armonk, NY: M.E. Sharpe.

———. 2003. Alternative strategic directions for the U.S. labor movement: Recent scholarship. *Labor Studies Journal* 28 (1): 133–55.

———. 2004. The effectiveness and limits of labor community coalitions: Evidence from south Florida. *Labor Studies Journal* 29 (1): 67–89.

Nussbaum, K. 2005. Letter to the editor of *Labor Talk*. *The Labor Educator*, June 1. http://www.laboreducator.org/ (accessed October 12, 2005).

NWU. *See* National Writers Union.

Nye, D. 1994. *American technological sublime*. Cambridge, MA: MIT Press.

O'Shaughnessy, K. C., and D. J. Flanagan. 1998. Determinants of layoff announcements following M&As: An empirical investigation. *Strategic Management Journal* 19 (10): 989–99.

Pacific Northwest Newspaper Guild. n.d. http://www.pnwguild.org/ (accessed June 8, 2007).

Palmer, B. D. 1992. *Working-class experience: Rethinking the history of Canadian labour, 1800–1991*. Toronto: McClelland & Stewart.

Pane, E. 1994. *Machine bites dog: A study of technology and work in the Ontario newspaper industry*. Toronto: Southern Ontario Newspaper Guild.

Pavlik, J. V., and E. E. Dennis, eds. 1993. *Media technology: A Freedom Forum Center reader*. Mountain View, CA: Mayfield Publishing.

Pellow, D. N., and G. Matthews. 2006. Immigrant workers in two eras. In *Challenging the chip*, ed. T. Smith, D. A. Sonnenfeld, and D. N. Pellow, 129–38. Philadelphia: Temple University Press.

Pellow, D. N., and L. S. Park. 2002. *The Silicon Valley of dreams: Environmental injustice, immigrant workers, and the high-tech global economy*. New York: New York University Press.

Perlman, S., and P. Taft. 1935. *History of labor in the United States*. New York: Macmillan.

Peters, J. W., and M. Maynard. 2005. The death of a GM dream. *International Herald Tribune*, December 1.

Peters, R., and T. Merrill. 1998. Clergy and religious persons' roles in organizing at O'Hare airport and St. Joseph Medical Center. In *Organizing to win: New research on union strategies*, ed. K. Bronfenbrenner, R. L. Seeber, and R. A. Oswald, 164–77. Ithaca, NY: ILR Press.

Peterson, E. 2002. Coming together: Promises and pitfalls of Minnesota's corporate accountability campaigns. *WorkingUSA* 5 (3): 46–80.

Peterson, L. 1983. The One Big Union in international perspective: Revolutionary industrial unionism, 1900–1925. In *Work, community and power: The experience of labor in Europe and America, 1900*–1925, ed. J. E. Cronin and C. Sirianni, 49–97. Philadelphia: Temple University Press.

Polaski, S. 2007. U.S. living standards in an era of globalization. *Carnegie Endowment for International Peace*, Policy Brief No. 53, July.

Porat, M. U. 1977. *The information economy: Definition and measurement*. Office of Telecommunications Special Publication 77-12, May. Washington, D.C.: U.S. Department of Commerce.

Porter, E. 2005. True or false: Outsourcing is a crisis. *New York Times*, June 19.

Powell, W. W., and K. Snellman. 2004. The knowledge economy. *Annual Review of Sociology* 30: 199–220.

Prindle, D. F. 1988. *The politics of glamour: Ideology and democracy in the Screen Actors Guild*. Madison: University of Wisconsin Press.

Professional Writers Association of Canada. 2006. Canadian professional writers survey. May. Toronto: Professional Writers Association of Canada.

PWAC. *See* Professional Writers Association of Canada.

Ragan, S. 2003. Corporate governance and social responsibility: Building the links. Union Network International. http://www.union-network.org/uniindep.nsf/0/0240de313e8f1a64c1256e5a0043fa88?OpenDocument (accessed July 1, 2006).

Rai, S. 2005a. A tutor half a world away, but close as a keyboard. *New York Times*, September 7. http://select.nytimes.com/ (accessed January 1, 2008).

——. 2005b. Indian concern to design IBM chips. *New York Times*, November 18. http://select.nytimes.com/ (accessed January 1, 2008).

——. 2006. India becoming crucial cog in the machine at I.B.M. *New York Times*, June 5.

Raines, J. C, L. E. Bersen, and D. M. Gracie, eds. 1982. *Community and capital in conflict: Plant closing and job loss*. Philadelphia: Temple University Press.

Reardon, R. 2008. Letter to members. *AFTRA Flash!* February 4. http://www.aftra.com/aftra/pres_page/08_02_04mem.html (accessed February 10, 2008).

Reich, R. B. 1991. *The work of nations: Preparing ourselves for 21st century capitalism*. New York: Vintage.

Reuters. 2007. India seeks higher-value outsourcing work. May 16. http://www.reuters.com/article/inDepthNews/idUSBOM10071920070516 (accessed September 5, 2007).

Reynolds, D., and J. Kern. 2001. Labor and the living-wage movement. *WorkingUSA* 5 (3): 17–45.

Rhee, N., and J. A. Sadler. 2007. Building an inclusive city. In *Labor in the new urban battlegrounds*, ed. L. Turner and D. B. Cornfield, 178–209. Ithaca, NY: ILR Press.

Rhode, G. 2003. Jobs with no frontiers—global mobility: A challenge to unions. UNI IBITS. http://www.union-network.org/UNIIBITS.nsf/ (accessed June 5, 2006).

Roach, S. S. 2005. The global labor arbitrage. In *Flying on one engine*, ed. T. R. Keene, 19–38. New York: Bloomberg Press.

Robb, D. 1999a. It's still two parts for actors. *The Hollywood Reporter*, January 29.

——. 1999b. SAG-AFTRA ballots due at 5 P.M. *The Hollywood Reporter*, January 25. http://hollywoodreporter.com/ (accessed March 16, 2006).

Roberts, D. 2005. Waking up to their rights. *Business Week*, August 22–29: 123–28.

Roberts, K., and K. Stahl. 2002. Jurisdiction fight heats up: Producers are battling actors in a jurisdiction fight. http://www.robertsstahl.com/articles/producers_jurisdiction.htm (March 20, 2007).

Robinson, I. 2002. Does neoliberal restructing promote social movement unionism? In *Unions in a globalized environment: Changing borders, organizational boundaries and social roles*, ed. B. Nissen, 189–235. Armonk, NY: M.E. Sharpe.

Rodino-Colocino, M. 2007. High-tech workers of the world, unionize! A case study of WashTech's "new model of unionism." In *Knowledge workers in the information society*, ed. C. McKercher and V. Mosco, 209–27. Lanham, MD: Lexington Books.

Rosa, J. 1999. DecisionOne averts union bid—management persuades employees to cancel vote prior to layoff. *Computer Reseller News*, February 8.

Ross, A. 2003. *No-collar: The humane workplace and its hidden costs*. New York: Basic Books.

Rothenburger, A. 1999. Microsoft hires temps permanently. *In These Times*, February 7.

Rubenstein, S. A., and T. A. Kochan. 2001. *Learning from Saturn: Possibilities for corporate governance and employee relations*. Ithaca, NY: ILR Press.

Rubin, G. 1975. The traffic in women: Notes on the 'political economy' of sex. In *Toward and anthropology of women*, ed. R.R. Reiter, 157–210. New York: Monthly Review Press.

Russial, J. 1989. Pagination and the newsroom: Great expectations. PhD diss., Temple University.

SAG. *See* Screen Actors Guild.

Saint-Simon, H., Comte de. 1952. *Selected writings*. Ed. and trans. F. M. H. Markham. Oxford: Basil Blackwell.

Salomon, A. 2007. AFTRA, IATSE develop closer relationship. *Backstage*, July 25.

Schiller, D. 1999. *Digital capitalism*. Cambridge, MA: MIT Press.

Schiller, H. I. 1973. *The mind managers*. Boston: Beacon Press.

Scott, D. T. 1987. *Technology and union survival: A study of the printing industry*. New York: Praeger.

Screen Actors Guild. 2003. Constitution and by-laws. Los Angeles: Screen Actors Guild.

——. 2004. SAG marks strong second year for global rule one protections. May 12. http://www.sag.org/ (accessed March 20, 2007).

——. 2006. Screen Actors Guild national board authorized member vote on dues increase. October 22. http://www.sag.org/ (accessed March 20, 2007).

——. 2007a. Constitution and by-laws. July. http://www.sag.org (accessed August 27, 2007).

——. 2007b. Statement from Screen Actors Guild, DGA and IATSE on tax ruling affecting J.O.B.S. Act production incentive. February 9. http://www.sag.org/ (accessed March 20, 2007).

——. 2007c. SAG history. http://www.sag.org/history/ (accessed August 27, 2007).

SEIU. *See* Service Employees International Union.

Sengupta, S. 2007. Electricity crisis hobbles an India eager to ascend. *New York Times*, May 21.

Service Employees International Union. n.d.-a http://www.purpleocean.org (accessed December 10, 2005).

——. n.d.-b Fast facts. http://www.seiu.org/about/fast_facts/index.cfm (accessed August 21, 2007).

Shniad, S. 2005. Drawing lessons from the TWU-Telus dispute. *Canadian Dimension*, December 1. http://canadiandimension.com/articles/2005/12/01/224/ (accessed December 1, 2005).

——. 2007. Neo-liberalism and its impact in the telecommunications industry: One trade unionist's perspective. In *Knowledge workers in the information society*, ed. C. McKercher and V. Mosco, 299–309. Lanham, MD: Lexington Books.

Shoch, J. 2001. Organized labor versus globalization: NAFTA, fast track, and PNTR with China. In *Rekindling the movement: Labor's quest for relevance in the 21st century*, ed. L. Turner, H. C. Katz, and R. W. Hurd, 275–313. Ithaca, NY: Industrial Labor Review Press.

Silicon Valley Toxics Coalition. 2008. http://www.etoxics.org/site/PageServer (accessed February 8, 2008).

Simon, L., Director of Legislative and Public Affairs for the Los Angeles local of the American Federation of Television and Radio Artists. 2006. Interview with authors. Los Angeles, February 6.

Sinha, P. 2004. Dilemma of organizing IT workers—the case of India. Paper presented to the Australian Labour Market Research Workshop, December 6–7, at the University of Western Australia, Perth.

Sleigh, S. R. 1998. *On deadline: Labor relations in newspaper publishing*. Putnam Valley, NY: Social Change Press.

Smith, R. 2007. Union touts second life virtual protest against IBM as a success. *WRAL Local Tech Wire*, November 14. http://www.localtechwire.com/business/local_tech_wire/opinion/blogpost/2040566/ (accessed January 4, 2008).

Smith, T., D. A. Sonnenfeld, and D. N. Pellow, eds. 2006. *Challenging the chip*. Philadelphia: Temple University Press.

Smythe, D. W. 1977. Communications: Blindspot of Western Marxism. *Canadian Journal of Political and Social Theory* 1 (3): 1–27.

Srinivasan, S. 2005. Indian workers resist union for outsourcers. *Journal Gazette*, January 25. http://global.factiva.com/ (accessed July 12, 2006).

Statistics Canada. 2005. Study: Diverging trends in unionization. *The Daily*, April 22. http://www.statcan.ca/Daily/English/050422/d050422c.htm (accessed May 15, 2007).

——. 2007. Unionization rates in the first half of 2006 and 2007. Perspectives on labor and income. August. http://www.statcan.ca/start.html (accessed January 26, 2008).

Stelter, B. 2007. Freelancers walk out at MTV networks. *The New York Times*, December 11.

Stone, K. V. W. 2004. *From widgets to digits: Employment regulation for the changing workplace*. Cambridge: Cambridge University Press.

Story, L. 2007. It's an ad, ad, ad, ad world. *New York Times*, August 6.

SVTC. *See* Silicon Valley Toxics Coalition.

Sweeny, J. 2001. Putting the guidelines to work. OECD Observer, May. http://www.oecdobserver.org/news/fullstory.php/aid/444/Putting_the_Guidelines_to_work.html (accessed February 25, 2008).

Sweeny, J., and K. Nussbaum. 1989. *Solutions for the new workforce: Policies for a new social contract*. Washington, D.C.: Seven Locks Press.

Swift, J. 2003. *Walking the union walk*. Ottawa: Communications, Energy and Paperworkers Union of Canada.

Tait, V. 2005. *Poor workers unions: Rebuilding labor from below*. Cambridge, UK: South End Press.

Tan, A. M. 2006. San Francisco launches California's first film and television rebate program. *San Francisco Sentinel*, April 6. http://www.SanFranciscoSentinel.com/ (accessed April 6, 2006).

Taras, D. G., J. T. Bennett, and A.M. Townsend, eds. 2004. *Information technology and the world of work*. New Brunswick, NJ: Transaction Publishers.

Taylor, K. 2005. Just who won at the CBC? *The Globe and Mail*, October 5.

Teena, K. 2006. Life begins at 30, with a ring from call centres. *The Times of India*, December 20. http://timesofindia.indiatimes.com/Cities/Mumbai/Life_begins_at_30_with_a_ring_from_call_centres/articleshow/856592.cms (accessed September 3, 2007).

Telegraph.co.uk. 2007. The new face of outsourcing. March 19. http://www.telegraph.co.uk/money/main.jhtml?xml=/money/2007/03/18/ccouts18.xml (accessed September 5, 2007).

Telus Corporation. 2006. Growing together. Corporation annual report. http://about.telus.com/investors/annualreport2006/ (accessed November 9, 2007).

Terlep, S. 2007. GM pushes job changes: Plants where UAW balks may lose new work. *Detroit News*, May 25.

Terranova, T. 2000. Free labor: Producing culture for the digital economy. *Social Text* 18 (2): 33–58.

——. 2004. *Network culture: Politics for the information age*. London: Pluto Press.

Time Warner. 2008 Corporate website. http://www.timewarner.com/corp/ (accessed January 15, 2008).

Timmons, H. 2004. Reuters plans to triple jobs at site in India. *New York Times*, October 8.

Tollefson, T. 2005. Indian trade unionists tour U.S. *People's World Weekly Newspaper Online*, January 20. http://www.pww.org/article/view/6364/1/248/ (accessed September 7, 2007).

Touraine. A. 1969/1971. *The post-industrial society—tomorrow's social history: Classes, conflicts and culture in the programmed society*. Trans. L. F. X. Mayhew. New York: Random House.

Tracy, J. F. 2006. "Labor's monkey wrench": Newsweekly coverage of the 1962–63 New York newspaper strike. *Canadian Journal of Communication* 31: 541–60.

Tracy, J. F., and M. L. Hayashi. 2007. A libratariat? Labor, technology, and librarianship in the information age. In *Knowledge workers in the information*, eds. C. McKercher and V. Mosco, 53–67. Lanham, MD: Lexington Books.

Traub-Merz, R., and J. Eckl. 2007. International trade union movement: Mergers and contradictions. International Trade Union Cooperation, briefing paper no.1. Bonn: Friedrich Ebert Foundation. Available at http://www.fes.de/gewerkschaften/english.html (accessed February 25, 2008).

Tremblay, D., R. Paquet, and E. Najem. 2006. Telework: A way to balance work and family or an increase in work-family conflict. *Canadian Journal of Communication* 31 (3): 715–31.

Trott, B. 2007. Immaterial labor and world order: An evaluation of a thesis. *Ephemera: Theory and Politics in Organization* 7 (1): 203–32.

Turner, L. 2004. Labor and global justice: Emerging reform coalitions in the world's only superpower. *Industrielle Beziehungen* 11 (1–2): 92–111.

Turner, L., and D. B. Cornfield, eds. 2007. *Labor in the new urban battlegrounds*. Ithaca, NY: ILR Press.

Turner, L., and R. W. Hurd. 2001. Building social movement unionism. In *Rekindling the movement: Labor's quest for relevance in the 21st century*, eds. L. Turner, H. C. Katz, and R. W. Hurd, 9–26. Ithaca, NY: Cornell University Press.

Turner, L., H. C. Katz, and R. W. Hurd, eds. 2001. *Rekindling the movement: Labor's quest for relevance in the 21st century*. Ithaca, NY: Cornell University Press.

UNCTAD. *See* United Nations Conference on Trade and Development.

UNI. *See* Union Network International.

Union Network International. 2003a. UNI Apro finance sector committee meeting. Report on UNI Apro finance sector activities, Hanoi, Vietnam, October 7–8. http://www.union-network.org/UNIApron.nsf/0/031f6833c221329ac1256fa4005c18c1/$FILE/UNI%20Apro%20Finance%20sector%20activities%20report.pdf (accessed July 11, 2006).

——. 2003b. Women and the information society. March 12. http://www.union-network.org/uniwomen.nsf/9548462b9349db27c125681100260673/931f1c9c5447 05e1c1256ce700324 81a?OpenDocument (accessed November 14, 2006).

——. 2004a. Culture, union constitutions impeding women. September 2, http://www .unionnetwork.org/uniwomen.nsf/9548462b9349db27c125681100260673/1e7c50f1 a4909ee7c1256f03003abbab?OpenDocument (accessed November 14, 2006).

——. 2004b. UNI telecoms for the long run. http://www.union-network.org/unitelecom .nsf/0/4095407d99e4d4ddc1256ea9004a70ce/$FILE/UNI%20Telecom% 20Regulatory%20Statement%20Jun04-en.pdf (accessed July 12, 2006).

——. 2004c. UNI charter on offshore outsourcing. http://www.union-network.org/ unisite/in_depth/outsourcingoffshoring/pdf/offshoring-en.pdf (accessed June 12, 2006).

——. 2004d. Global call centre news. April. http://www.union-network.org/UNIsite/ Events/Campaigns/Callcenter/CallCentreNews2EnLoRes.pdf (accessed August 17, 2007).

——. 2005a. Outsourcing and contracting out: Failing to deliver quality services, secure jobs, and a sustainable telecom industry. http://www.union-network.org/uniapron .nsf/db1cb10297249246c1256f6c005f9629/0978058ebdc9a485c1257035003fb65c/ $FILE/6.%20Outsourcing%20and%20Contracting%20Out%20in%20the% 20Telecom%20Industry.pdf (accessed June 13, 2006).

——. 2005b. Make poverty history. December 6. http://www.union-network.org/ uniindep.nsf/GlobalCallAgainstPovertyCampaign?openpage (accessed 12 July 2006).

——. 2005c. Call to action against AIDS in Africa. July 3. http://www.union-network .org/uniafrican.nsf/0/FDB79D996003 1C9FC12571A000459EC1?OpenDocument (accessed July 12, 2006).

——. 2005d. UNI call centre charter. http:www.union-network.org/UNIsite/Events/ Campaigns/CallCenter/CCCharterLoResEng.pdf (accessed June 12, 2006).

——. 2005e. Women's work: Organizing for the future. 2nd UNI World Women's Conference, August 20–21, Chicago.

——. 2005f. UNI passport. http://www.union-network.org/UNISITE/Groups/PMS/ Passport/LeafletEng.pdf (accessed July 12, 2006).

——. 2005g. For gender equality to become a reality in French telecommunications companies. November 2. http://www.union-network.org/uniwomen.nsf/9548462b 9349db27c125681100260673/7d5794698403ff79c12570ad002cf071? OpenDocument (accessed Novemebr 14, 2006).

——. 2005h. *News Flash*. April 7. http:www.uniglobalunion.org (accessed October 19, 2007).

——. 2005i. Stepping up fight for equality and to end discrimination. August 21. http://www.union-network.org/uniwomen.nsf/9548462b9349db27c125681100 260673/f921b944c9c7769ac1257065004d6196?OpenDocument (accessed Novemebr 14, 2006).

——. 2006a. Message from Barbara Easterling, UNI World Women's President. March 8, http://www.union-network.org/uniwomen.nsf/9548462b9349db27c125 681100260673/c563d0246aedda6cc125712a0051bc76?OpenDocument (accessed November 14, 2006).

——. 2006b. Opening doors for unions: Global agreements. http://www.union-network.org/uniflashes.nsf/e6b6391ed587d2d6c12571c3002fd064/$FILE/UNIRepGlobAgree-En.pdf (accessed June 4, 2006).

——. 2006c. Telefónica under pressure over Puerto Rico events. June 15. http://www.union-network.org/unitelecom.nsf/d4b55a8f47df3454c125701a00333cfa/c6d452a9c1ea5695c125718e003954e4?OpenDocument (accessed June 22, 2006).

——. 2006d. UNI-Apro co-sponsors Asia Pacific cooperative conference. October 7. http://www.union-network.org/uniapron.nsf/aa46e8f772bb0978c1256f6d00415cc7/abaf52810c289c64c12571b60026abfc?OpenDocument (accessed June 4, 2006).

——. 2006e. UNI telecom seeking respect for Atento Puerto Rico union rights. October 31. http://www.union-network.org/unitelecom.nsf/1eec6d83918954ebc125701a0032807b/f2c679c844940f27c1257218000e0843?OpenDocument (accessed June 4, 2007).

——. 2006f. UNITES India signs first agreement. June 26. http://www.union-network.org/uniflashes.nsf/0/D44C7952B0EEBE4CC125719C003508A2?OpenDocument (accessed July 11, 2006).

——.2007a. UNI Telefónica global union tells Telefónica "no to precarious work." January 9. http://www.union-network.org/unitelecom.nsf/35a9b6bab9eedd74c125701400504b36/5b4430782cc36545c12572680076fa9d?OpenDocument (accessed June 4, 2007).

——. 2007b. UNI global union and Quebecor World reach global agreement on labour rights. May 5. http://www.union-network.org/UNIAmericas.nsf/0/1B5ECC5A64962D90C12572D1002BEA78?OpenDocument (accessed June 4, 2007).

——. 2007c. Global framework agreements. http://www.union-network.org/UNIsite/In_Depth/Multinationals/GFAs.html (accessed January 3, 2008).

——. n.d.-a. About us: Press pack. http://www.uniglobalunion.org/UNIsite/About_Us/About_Us.html (accessed February 25, 2008).

——. n.d.-b. Call centres. http://www.union-network.org/uniindep.nsf/callcentres?openpage (accessed July 1, 2006).

——. n.d.-c. Call centre newsletters. http://www.union-network.org/UNIsite/Events/Campaigns/CallCentreNewsletters.html (accessed July 1, 2006).

United Kingdom. 1968. *Royal commission on trade unions and employers' associations*. Cmnd. 3623. London: Her Majesty's Stationary Office.

United Nations Conference on Trade and Development. 2004. *World investment report: The shift towards services*. Geneva: United Nations Conference on Trade and Development.

United Steelworkers. 2007. USW, Amicus take first steps toward global super union. News release, April 18. http://www.uswa.org/uswa/program/content/3936.php (accessed September 13, 2007).

Upadhya, C., and A. R. Vasavi, eds. 2008. *In an outpost of the global economy: Work and workers in India's information technology industry.* New Delhi: Routledge.

Ursell, G. 1998. Labour flexibility in the UK commercial television sector. *Media Culture & Society* 20 (1): 129–53.

U.S. Bureau of Labor Statistics. 2007. *Union members summary*. USDL 07-0113. January. Washington, D.C.: Department of Labor.

———. 2008. Union members in 2007. USDL 08-0092. January 25. Washington, D.C.: Department of Labor.

U.S. Newswire. 2005. Global unionism rising to challenge global economy: Background information on UNI congress, convention. *U.S. Newswire*, August 7. http://global.factiva.com/ (accessed July 14, 2006).

USW. See United Steelworkers.

Van Jaarsveld, D. D. 2004. Collective representation among high-tech workers at Microsoft and beyond: Lessons from WashTech/CWA. *Industrial Relations* 43 (2): 364–85.

Vlessing, E. 2004. Canada launches attack in runaway prod'n war. *The Hollywood Reporter*, October 14. Retrieved from http://www.friends.ca/ (accessed April 5, 2006).

Vonnegut, K. 1963. *Cat's cradle*. New York: Dell.

Vosko, L. 2000. *Temporary work: The gendered rise of a precarious employment relationship*. Toronto: University of Toronto Press.

———. 2006. Precarious employment: Towards an improved understanding of labor market insecurity. In *Precarious employment: Understanding labor market insecurity in Canada*, ed. L. Vosko, 3–39. Montreal: McGill-Queen's University Press.

Voss, K. 1993. *The making of American exceptionalism: The Knights of Labor and class formation in the nineteenth century*. Ithaca, NY: Cornell University Press.

Walsh, L. D. 1985. A Study of the proposed merger of the International Typographical Union and the Newspaper Guild: 1974–1983. MSc thesis, Cornell University.

Walt Disney Company. 2008 Corporate website. http://corporate.disney.go.com/corporate/overview.html (accessed January 15, 2008).

Waring, M. 1988. *If women counted: A new feminist economics*. New York: Harper Collins.

———. 2003. Counting for something! Recognizing women's contribution to the global economy through alternative accounting systems. *Gender and Development* 11 (1): 35–43.

WashTech. n.d. http://www.washtech.org/ (accessed June 11, 2007).

———. 2005. Offshore tracker update. *WashTech News*, August. http://www.itpaa.org/modules.php?name=News&file=article&sid=1913 (accessed January 1, 2008).

Wasko, J. 2004. *How Hollywood works*. London: Sage.

Waterman, P., and J. Timms. 2004/2005. Trade union internationalism and a global civil society in the making. In *Global civil society*, ed. H. Anheier, M. Glasius, and M. Kaldor, 175–202. London: Sage Publications.

Webb, K. 2007. Rising dollar worries local film producers. *The Province*, August 2.

Weir, R. E. 1996. *Beyond labor's veil: The culture of the Knights of Labor*. University Park: Pennsylvania State University Press.

White, J. 2005. Losing Canadian culture: The danger of foreign ownership of telecom. *Canadian Centre for Policy Alternatives: Trade and Investment*, October 31. http://policyalternatives.ca/documents/National_Office_Pubs/2005/brief6_3_Losing_Canadian_Culture.pdf (accessed August 27, 2007).

Wigfield, A. 2001. *Post-Fordism, gender and work*. Aldershot, UK: Ashgate Publishing.

Winseck, D. 1998. *Reconvergence: A political economy of telecommunications in Canada*. Creskill, NJ: Hampton Press.

Winter, J. P. 2005. *Lies the media tell us*. Montreal: Black Rose Books.

Wong, C. 2005. Telus Corp. announces merger of Telus Mobility with wireline division. *CBC.ca*, November 24. http://www.cbc.ca/ (accessed July 8, 2007).

Woods, A. 1963. *Modern newspaper production*. New York: Harper.

Woolgar, S. 1996. Technologies as cultural artifacts. In *Information and communications technologies: Visions and realities*, ed. W. H. Dutton, 87–101. Oxford: Oxford University Press.

Wright, B. D., ed. 1987. *Women, work and technology*. Ann Arbor: University of Michigan.

Zerker, S. F. 1982. *The rise and fall of the Toronto Typographical Union 1832–1972: A case study of foreign domination*. Toronto: University of Toronto Press.

Zhao, Y., and R. Duffy. 2007. Short-circuited? The communication of labor struggles in China. In *Knowledge workers in the information society*, ed. C. McKercher and V. Mosco, 229–47. Lanham, MD: Lexington Books.

Zimbalist, A. 1979. Technology and the labour process in the printing industry. In *Case Studies on the Labor Process*, ed. A. Zimbalist, 103–26. New York: Monthly Review Press.

Zuboff, S. 1988. *In the age of the smart machine*. New York: Basic Books.

Index

AAAA. *See* Associated Actors and Artistes of America
AAUP. *See* American Association of University Professors
ABC television, 5, 39, 108
Abitibi, 111
Academy of Television Arts & Sciences, 202–3
Actors' Equity, 103, 127, 135
Actors Working for an Actors Guild, 131
ACTRA. *See* Alliance of Canadian Cinema, Television and Radio Artists
AFL-CIO. *See* American Federation of Labor and Congress of Industrial Organizations
AFTRA. *See* American Federation of Television and Radio Artists
AGT. *See* Alberta Government Telephones Commission
Alberta Government Telephones Commission, 140. *See also* Telus
Alliance@IBM, 116, 172, 175, 178
Alliance of Canadian Cinema, Television and Radio Artists, 103, 205
Alliance of Independent Video and Filmmakers, 202–3
Alliance of Motion Picture and Television Producers, 135, 137
Amazon.com, 116, 170
Amber, Arnold, 121, 124
American Association of University Professors, 50–51; joint conference with Newspaper Guild, 51
American Economics Association, 2
American Express, 190
American Federation of Labor, 96, 99, 101, 102, 128, 164, 204
American Federation of Labor and Congress of Industrial Organizations, 1, 9–10, 42, 72, 73, 76, 102, 113, 136, 159, 162
American Federation of Television and Radio Artists, xi, 17–18, 103, 114, 125–39, 147–48, 199–200, 220; founding and origins of, 128; joint negotiations with SAG, 125, 127, 129, 135, 136; merger attempts, 125, 129, 130–34, 199, 205–7; restructuring, 135; and runaway production, 202–3, 205; shared membership with SAG, 127, 205–6
American Jobs Creation Act, 203
Amicus, 46, 209
AOL, 4, 40, 189. *See also* Time Warner
Apted, Michael, 203
Arar, Maher, vii–ix
Asian Immigrant Women's Advocates, 175

Asner, Ed, 129, 131
Associated Actors and Artistes of America, 128, 136
Association of Independent Commercial Producers, 202–3
Atento, 214. *See also* Telefónica
AT&T, 5, 39, 108, 132, 157–58
audience labor, 62
autonomous Marxism, 16, 36–37, 66–69, 79; and social factory, 67–68; and worker identity, 66–67

Babbage, Charles, 14, 29
BBC, 122
BCE. *See* Bell Canada Enterprises
BC Tel, 18, 140–1, 143, 146. *See also* Telus
Bell telephone companies, U.S., 103, 108
Bell Canada Enterprises, 5, 6, 40, 110, 111, 149; and pay equity, 73, 111
Bell, Daniel, 14, 33–35, 64; on changing American occupational structure, 33–35; on cultural contradictions of capitalism, 34
Blain, Mike, 166
Braverman, Harry, 14, 35, 85, 123; on control over the labor process, 35
British Telecom, 44–45
broadcasting, 4, 6, 11, 17, 39–40, 48–49, 55, 62, 71, 108, 119, 122–3, 126, 128–9
BT. *See* British Telecom
Burger, Anna, 72
business unionism, 52, 151–55, 177; definition of, 153; relationship to psychological contract, 154–5

Calgary Herald, 158
call centers, 3, 11, 23, 25–26, 31, 54, 63, 69, 75, 108, 117, 145, 181, 184, 186, 188, 192, 214, 216–18; role in occupational hierarchy, 25–26; development of Call Centre Charter, 217–8
Campaign for Responsible Technology, 174
Canada Industrial Relations Board, 109–110, 119–120, 142–45, 147; Island Telecom decision, 143; role in consolidation of CBC bargaining units, 120; role in TWU-Telus dispute, 143–45 support for labor convergence, 109–10;
Canadian Association of University Teachers 50–52; joint conference with CEP, 51–52
Canadian Auto Workers, 103, 109, 113, 147, 165
Canadian Broadcasting Corp., 17, 43, 81–82, 95, 106, 108, 112, 119–26, 139, 144, 147–49; 2005 lockout, 17, 119, 122–24, 148; labor convergence at the CBC, 81–2
Canadian Centre for Policy Alternatives, 111
Canadian Film and Television Production Association, 204–5
Canadian Freelance Union, 73–74, 176, 178, 179; organizing drive, 176
Canadian Human Rights Commission, 73
Canadian Labour Congress, 73, 102
Canadian Media Guild, ix, 81–82, 119–25. *See also* Communications Workers of America
Canadian Union of Public Employees, 119
Canwest Mediaworks, 5, 40, 112, 145
Career Launcher, 189–90
Castells, Manuel, 14, 28–29, 64
Caucus of Artists for Merger, 130–31
Caucus for Television Producers, Writers and Directors, 202–3
CAW. *See* Canadian Auto Workers
CBC. See Canadian Broadcasting Corp.
"CBC Unlocked," 123–24
CBS, 6
Center on Wisconsin Strategy, 172
Centre for Business Process Outsourcing Professionals, 217

CEP. *See* Communications, Energy and Paperworkers Union of Canada
CFTPA. *See* Canadian Film and Television Production Association
CFU. *See* Canadian Freelance Union
Change to Win, 10, 42, 72; break from AFL-CIO, 10
China, 2, 12, 22, 97, 181, 183, 187, 190–92, 194, 197–8, 208; outsourcing to, 190–92
Cingular 116
CIRB. *See* Canada Industrial Relations Board
Cirque du Soleil model, 185
Civic Service Union, 141
class-based unionism, 82–83, 95–101; definition and types of, 82–83
Clear Channel, 132–33
Clearnet Communications, 140, 143, 144
Cleaver, Harry, 66–67
CMG. *See* Canadian Media Guild
Coalition of Canadian Audio-Visual Unions, 113
Cold Mountain, 204
Colgate-Palmolive, 183–84
Communications, Energy and Paperworkers Union of Canada, ix, 11, 25, 27, 42–43, 56, 103, 105–6, 112, 113, 114, 141, 143, 145, 146, 147–49, 165, 176, 178–79, 210; and the CBC, 81–82, 119–125, 144; and convergence, 110–11; and freelancers, 73–74; national nature of, 42–43, 71–72, 103, 105, 109; and pay equity, 73, 111; political activity of, 111; and women, 71–74
Communications International. *See* Union Network International
Communications Workers of America, ix, 17, 19, 25, 27, 31, 42, 43, 51, 71–74, 84, 103, 105–6, 107–9, 134, 140–41, 146, 147–49, 156, 157, 163, 167–70, 172, 175, 178–79, 210, 214; and CBC bargaining unit consolidation, 81–82, 107, 112, 120–26, 140, 144, 148; and convergence, 115–25; customer service workers, 116–17; CWA Canada, xi; and outsourcing, 117, 184–85; political activity, 117–18. *See also* Newspaper Guild, Washington Alliance of Technology Workers
Communications Workers of Canada, 103, 109
computers, ix, 3–4, 11, 15, 18, 24–25, 29, 31–32, 37, 39–40, 47–48, 56, 63–65, 68–69, 72, 78, 91, 93, 109, 115, 152, 170, 174–5, 187–190, 194–5, 197
Congress of International Organizations, 102, 103
Connolly, John, 135, 136
Consumer labor. *See* voluntary labor
convergence: and commodification, 38–39; and corporations, 4–6, 39–40, 132–33, 139–47; definitions of, 23, 37–38, 40–41; and labor, 8, 9, 11–13, 17–18, 26, 39, 41–46, 82, 102, 108–15, 147–49; and technology, 3–4, 6, 37–39; and Washington Alliance of Technology Workers, 165–172. *See also* trade union convergence *See also* worker associations
Courtney, Marcus, 166, 169, 170, 179
COWS. *See* Center on Wisconsin Strategy
craft unionism, 2, 16–17, 39, 50–51, 82, 83–84, 85, 87–89, 94–95, 96, 97, 99, 101, 104–5, 126–27, 131, 138, 148, 168; and the ITU, 87–89, 104–5; and the Screen Actors Guild, 126;
Creators Federation, 179
CTVglobemedia, 5, 6
CWA. *See* Communications Workers of America

Daniels, William, 131–32, 133
Davis, Mike, 196–7
Deloitte Touche, 198
Denning, Michael, 21, 102

deskilling, 4, 14, 35, 54, 66, 68, 85, 107, 123, 147, 152–54, 220. *See also* Harry Braverman
DGA. *See* Directors Guild of America
digital capitalism, 23, 28, 77
Digitas, 190
Directors Guild of America, 103, 130, 135, 137, 201–3, 205. *See also* runaway production
Directors Guild of Canada, 202
Disney, 5, 39
division of labor, international 19, 181–2
domestic labor, 56–62; economic value of, 57; gender disparities in, 61; and household technology, 58–61; and scientific management, 59
Dow Jones, 108, 115
Drucker, Peter, 24
Dunn, Mathis L., Jr., 125
Dyer-Witheford, Nick, 14, 36, 69

Easterling, Barbara, 75
Economic Policy Institute, 117
EDS, 188
ED TEL, 140. *See also* Telus
Eisenstein, Zillah, 56, 62, 77
end of geography, 197–8
Estevenin, Karen, 168

Fair, Pamm, 133, 136, 203–4, 207
Federation of Telephone Workers of British Columbia. *See* Telecommunications Workers Union
feminist theory, 14; and political economy, 14, 53–76
FIET. *See* Union Network International
film, 3, 5, 17, 19, 22, 38–39, 114, 125–30, 135, 137, 181, 199–207, 220
Film and Television Action Committee, 138, 204–5, 206–7
Florida, Richard, 24–26, 30
Foley, Linda, 74
Fourastié, Jean, 14, 29
Freelancers Union, 175–6, 179
FTAC. *See* Film and Television Action Committee

G8, 8
Gates, Bill, 8, 47, 63, 142, 195
GCIU. *See* Graphic Communications International Union
GE, 5, 39, 60, 188, 198
General Motors, 156–7, 190
Gilbert, Melissa, 132, 139
Google, 5, 39
Graphic Artists Guild, 176
Graphic Communications International Union, 71–72, 84, 105, 109; relationship to the International Brotherhood of Teamsters, 71–72
Great Depression, 6–7, 21, 89, 102
Growing Stars, 189–190

H-1B visa, 171, 185–6, 195–6; role of Indian companies, 196
Haraway, Donna, 68, 79
Hardt, Michael, 29, 36–37, 66
Harper, Valerie, 132
health and safety, 63–65
Heston, Charlton, 129
Hewlett-Packard, 188, 190
HIT Entertainment, 183–84
Hollywood Post Alliance, 202–3
Hughes, 188
Huws, Ursula, 14, 35, 59–60, 61–62, 191

IATSE. *See* International Alliance of Theatrical Stage Employees
IBEW. *See* International Brotherhood of Electrical Workers
IBM, 11, 116, 188, 195, 218
ICICI Group, 193, 197; North American holdings of, 193, 197
IFJ. *See* International Federation of Journalists
IG Metall, 46
illegal immigrant labor, 63–64
ILO. *See* International Labour Organization
immaterial labor, 36–37, 66–69; and emancipatory potential of technology, 66

Independent Film and Television Alliance, 202–3
India, 6, 11, 22, 75, 117, 170–1, 181–2, 187–98, 208–9, 215–17, 219; infrastructure needs, 182; role in outsourcing, 187–98; UNI activism in, 215–17
industrial unionism, 42, 102–6
Industrial Workers of the World, 16, 81, 83, 95, 99–101, 104, 164; founding and origins of, 99–100; and "one big union," 81, 95; strikes, 100–110; suppression of, 100–101
Infocrossing, 195
information society, 14, 22, 23–25; definitions of, 27–29; theories of, 27–29, 33–37
Information Technology Association of America, 186
Infosys Technologies, 193–4, 195, 196; global back office system, 194
International Alliance of Theatrical Stage Employees, 103, 135, 202–3
International Association of Machinists and Aerospace Workers, 46, 156
International Brotherhood of Electrical Workers, 103, 141, 144
International Brotherhood of Teamsters, 9, 31, 71–72, 84, 109, 202–3
International Confederation of Free Trade Unions, 209–210
International Federation of Journalists, 210
International Graphical Federation. *See* Union Network International
International Labour Organization, 70, 77, 134, 212, 213, 215
International Trade Union Confederation, 209–10
International Typographical Union, 9, 16, 43, 51, 71–72, 82–95, 99, 102, 103, 104–5; and Communications Workers of America, 108; craft control by, 85, 88–95; departure of crafts from, 86, 87–88; founding and origins of, 84, 87; and inter-union rivalry, 87, 92–93; merger efforts by, 84, 93–94, 95; and strikes, 91–93; and technological change, 86–87, 90–95
International Union, United Automobile, Aerospace and Agricultural Implement Workers of America, 109, 156–7, 178
iPod, 47, 136
Island Telecom, 143
Island Telecom Advanced Solutions, 143
IT Professionals Forum, 209, 216
ITAS. *See* Island Telecom Advanced Solutions
ITU. *See* International Typographical Union
ITUC. *See* International Trade Union Confederation
IWW. *See* Industrial Workers of the World

Jennings, Philip, 211
Jobs With Justice, 117, 162, 163
journalism, 4, 20, 22, 40, 71, 108, 110, 121, 158, 181, 210

Kimbrough, Matt, 133–4, 137, 138, 207, 220
King County Labor Council, 166–7
Knights of Labor, 16, 42, 83, 95–99, 100, 101, 102, 104–5; and American Federation of Labor, 99; collapse of, 98–99; expansion of, 97–98; founding and origins of, 96–97; and Haymarket incident, 98; and strikes, 98
knowledge labor: definitions of, 23–27; outsourcing of, 36, 46; significance of, 45–52; and skill, 35–36, 54; theories of, 29–37; and women, 54–56, 76–80
Kohl, George, 116, 179
Kyoto Accord, 110

labor councils, 161–2
Labor Immigrant Organizing Network, 175
Lareau, Lise, 81, 125

legislation, U.S., 10, 44
Lem, Gail, 74
Leo Burnett, 190
London Free Press, 158

Machlup, Fritz, 14, 29–31, 32, 33
Marx, Karl, 14
Masur, Richard, 131
McGraw-Hill, 191
McKercher, Catherine, 20–22, 25, 37, 39, 41, 42, 85, 93–94, 107, 112, 126, 158, 169
McWilliams, Carey, 3
Meany, George, 9–10
Media and Entertainment International, 211. *See also* Union Network International
Membership First, 134
Metroland West Media Group, 189
Microsoft, 5, 11, 12, 18, 39, 115, 116, 152, 165–67, 168, 169–70, 171, 176, 187, 195–97; role in H1-B visa program; Tax Saver group, 170
Morgan, Ron, 134, 138
Mosco, Vincent, ix, 4, 20–22, 24, 38, 39, 40, 48, 56, 111, 112, 142, 146, 197
Motion Picture Association of America, 202–3
MTV, 176
Murdoch, Peter, 179

NAACP, 163
NABET. *See* National Association of Broadcast Employees and Technicians
NACU. *See* National Alliance of Communication Unions
National Alliance of Communication Unions, 113, 114
National Association of Broadcast Employees and Technicians, 71–72, 105, 108, 109, 119, 120
National Association of Working Women, 74
National Federation of Telephone Workers, 103
National Organization of Women, 162
National Writers Union, 167, 176, 178
NBC television, 5, 39, 108, 198
Negri, Antonio, 29, 36–37, 66
neo-liberalism, 2, 7–9, 10, 18, 36, 44, 52, 110, 140, 152, 154, 164, 182, 208, 220; definition of, 7; international growth of, 8
network society, 23, 28, 126
New Deal, 44, 102, 127
New Democratic Party, 159, 165
New Trade Union Initiative, 117
New York Daily News, 91
New York Times, 108
New York World Telegram, 92
News Corp., 5, 39
Newspaper Guild, 43, 51–52, 71–72, 74, 95, 103, 105, 109, 130; and International Typographical Union, 84, 92; as sector of Communications Workers of America, 108, 115, 168–9
Nissen, Bruce, 156–7, 159, 161–65, 178
Nonde, Joyce, 73
Nussbaum, Karen, 74, 76
NWU. *See* National Writers Union

One Big Union, 81, 83, 101, 104
O'Neill, Juliet, viii–ix
OpenNet Initiative, 145
outsourcing, 7, 18, 19, 20, 36, 54–55, 114, 147, 168, 170–71, 179, 182–221; extent and growth of, 184–91, 219–21; from developing to developed world, 192–97; and H1–B visas, 189–90, 195–6; knowledge work, 187, 189–91, 193; manufacturing work, 181, 183–4; media work, 181, 189–191; service work, 181, 183, 188; to the developing world, 186, 187–191; within the developed world, 191–2; within national borders, 186; worker

resistance to, 182, 208–17. *See also* runaway production, Union Network International.

Pacific Newspaper Group, 43, 112, 145
Park, Lisa Sun-hee, 15, 26, 63–66, 172–75
pay equity, 55, 77
Pellow, David Naguib, 15, 26, 63–66, 172–75
Performers Alliance, 131
Perot Systems, 188
Porat, Marc Uri, 14, 31–33
post-industrial society, theories of, 33–37, 54, 64; and women, 54–55, 76–79
precarious employment, 12, 54, 154–5, 165–79
Pressmart Media Ltd., 189
Procter & Gamble, 190
Producers Guild of America, 202–3
product recalls, 183–84
Professional Writers Association of Canada, 73–74, 176
Publicis, 190
PWAC. *See* Professional Writers Association of Canada

al Qaeda, vii–ix
Quebecor, 6, 40, 74, 176, 215
QuébecTel, 140

radio, 4, 21, 48, 62, 72, 102–3, 119–121, 123–4, 127–8, 132, 136, 142, 192, 205
Radio Canada, 120
RC2 Corporation, 183–4
Reagan, Ronald, 7, 60, 129
Reardon, Roberta, 135, 136, 137
Reich, Robert, 24–25
Reskilling, 4, 35, 54, 85, 87
Restore Respect, 132
Reuters, 189
Rodino-Colocino, Michelle, 171
Rogers Communications Inc., 5, 40
Rosenberg, Alan, 134–35
runaway production, 19, 137–8, 189, 199–207, 220; and Canada, 201–2, 206–7; incentives to prevent, 203–4; Monitor Report on, 201–2
Runaway Production Alliance, 137–8, 202–4, 205

Saatchi & Saatchi, 190
SAG. *See* Screen Actors Guild
Saint-Simon, Comte de, 14, 29
Santa Clara Center for Occupational Safety and Health, 174
Santa Clara County. *See* Silicon Valley
Saturn, 156–7
Save Sag, 131, 133
Schiller, Daniel, 28
Schiller, Herbert, 14, 35
SCCOSH. *See* Santa Clara Center for Occupational Safety and Health
Screen Actors Guild, ix, 17–18, 103, 114, 125–139, 147–8, 199–207, 220; founding and origins of, 127–8; and global rule one, 134, 137–8, 204–5, 207, 220; joint negotiations with AFTRA, 125, 127, 129, 135, 136; merger attempts, 125, 129, 130–34, 199, 205–6; political parties in, 129–31; and runaway production, 137–8, 199–207; shared membership with AFTRA, 127, 205–6
Screen Directors Guild. *See* Directors Guild of America
Screen Extras Guild, 131
Screen Writers Guild, 127
SEIU. *See* Service Employees International Union
Service Employees International Union, 10, 72, 163–4, 165, 185; break with AFL-CIO, 72; and Justice for Janitors, 163; membership, 163; raiding charges, 165
Shaw Communications Inc., 40
Silicon Valley, 15, 18, 25, 62–65, 66, 78, 152, 165, 172–75, 179, 185

Silicon Valley Toxics Coalition, 174
Simon Fraser University, 195
Simon, Leslie, 134
Sindicato dos Trabalhadores em Telecomunicações, 214, 217–8
SINTETEL. *See* Sindicato dos Trabalhadores em Telecomunicações
Smythe, Dallas, 21, 62
social contract, postwar 7–8, 36
social movement unionism, 16, 18, 19, 52, 70–71, 74, 76, 79, 95, 111–2, 152, 158–165, 167, 172, 174, 175, 177–8, 179, 181, 199, 209, 218; definition of, 18; ideal type, 160–1; relationship to UNI, 74–75; relationship to value-added unionism, 152; and SEIU, 162
Socialist Party of Canada, 101
South Bay Labor Council, 175
Southern Christian Leadership Conference, 162
Sprint, 108
Starbucks, 191
Starcom, 190
Stern, Andrew, 185
Supreme Court of Canada, 73
Supreme Court of U.S., 134
surveillance, 3, 6, 8, 25, 41, 68, 111, 217
SVTC. *See* Silicon Valley Toxics Coalition

Taft-Hartley Act, 89
Target stores, 181
Tata Consultancy Services, 194–97; global expansion, 194; role in H1-B visa program, 195–6
Teamsters. *See* International Brotherhood of Teamsters
Technological change, 2–4, 86–87, 90–95, 128–9, 132, 136–7
TechsUnite.org, 171, 172
telecommunications, 6, 11, 18, 20, 26, 37, 40, 43–44, 48, 55, 65, 71–72, 74–75, 78, 103–4, 107–15, 139–44, 146, 148–9, 167, 169, 181, 188, 210–11, 213, 215
Telecommunications Workers Union, ix, 18, 103, 107–8, 112, 114, 139–47, 188 and 2005 lockout, 140, 145–6; bargaining unit consolidation, 109–10, 112; and convergence, 142–45; founding and origins of, 140–1; strikes, 141
Telefónica, 213–15, 216
Telemundo, 198
television, 5–6, 17, 32, 39, 47–48, 55, 60, 62, 71, 103–4, 108, 110, 114, 119–22, 124–29, 131–2, 135, 137–39, 146, 190, 198–201, 203, 205, 207, 220
Television Authority, 128–9
Telus, 18, 40, 107–8, 112, 139–47, 188; and convergence, 141–44; expansion of, 139–40, 141
Telus Mobility, 143
Thomas the Tank Engine, 183
Time Warner, 4, 39, 40, 132, 189
Torstar Corp., 189; Metroland, 189; *Toronto Star*, 91–92;
toxic materials, 63, 64; and Superfund sites, 63
trade union convergence, 11–13, 18–19, 20, 39, 41–46, 74–75, 76, 80, 107–49, 181; bargaining unit consolidation, 17, 112, 119–25; coalitions, 113, 114, 147, 205–7; international federations, 113, 207–19; mergers efforts, 17, 42, 46, 71–72, 74, 106, 112, 125–39; and national unions, 109, 165
trade union membership: Canada and U.S., 1–2; and wage premium, 73; and women, 72–73
trade union membership decline, reasons for 2–10, 151–55. *See also* neo-liberalism, technological change, convergence
Trades and Labour Congress, 99, 101
Transport and General Workers Union, 46

TVA. *See* Television Authority
TWU. *See* Telecommunications Workers Union

UAW. *See* International Union, United Automobile, Aerospace and Agricultural Implement Workers of America, 109
UNCTAD. *See* United Nations Conference on Trade and Development
UNI. *See* Union Network International
Union for Information Technology Enabled Services Professionals, 217
Union of Needletrades, Industrial and Textile Employees, 156
Union Network International, 19–20, 42, 71, 113–4, 199, 207–19 Call Centre Charter, 217–8; founding and origins of, 19–20, 211; Global Framework Agreements, 212–15, 216, 221; and global labor solidarity, 216–19; media sector, 211; and outsourcing, 215–16; and professionals, 212; telecom sector, 113, 211; UNI Africa, 73; and women, 74–75, 211–12; and youth, 212
UNISON, 70
Unite, 46
United Electrical Workers, 103, 162–3
United Food and Commercial Workers, 163
United Mineworkers, 162–3
United Nations Conference on Trade and Development
United Nations Fourth World Conference on Women, 70
United Nations System of National Accounts, 57
United Parcel Service, 23, 31
United States National Labor Relations Board, 214
United Steelworkers, 45–46, 209
UNITES. *See* Union for Information Technology Enabled Services Professionals
UNSNA. *See* United Nations System of National Accounts
upskilling, 4, 35, 54, 85
U.S. Citizenship and Immigration Service, 63, 79
USW. *See* United Steelworkers

value added unionism, 18, 152, 155–58
Vancouver Province, 112
Vancouver Sun, 112
Viacom, 6, 176
Voices for Change, 145
voluntary labor, 39, 62, 68, 79

Wagner Act, 102, 103, 140
Wal-Mart, 163, 181, 191
war on terror, 8
Waring, Marilyn, 57
Warner Brothers, 139
Washington Alliance of Technology Workers, ix, 12, 18, 31, 71, 116, 152, 165–72, 178, 179, 184–5, 188, 189–90, 209, 217; founding and origins of, 166–7; and Offshore Tracker, 171; organizing efforts of, 169–70; political activities of, 170–72, 184; worker assistance, 170, 172. *See also* Communications Workers of America
Washington Post, 108
WashTech. *See* Washington Alliance of Technology Workers
Wasko, Janet, 20
Watts, Glenn, 157
Wells Fargo, 191
Western Electric, 103
Western Federation of Miners, 99
WGA. *See* Writers Guild of America
Winnipeg General Strike, 81, 101
Wipro Technologies, 194–96; global outsourcing offices, 194
women, 15–16, 53–79; and household labor, 56–61, 78; and labor saving device thesis, 59–60; and trade unions, 9–10, 69–76; and wage gap,

55–56; and wage labor in the home, 62–66, 78; and workforce, 9
worker associations, 11–13, 75–76, 165–79. *See also* social movement unionism
Working America, 71, 76
Working Partnerships USA, 175
Working Today, 71, 172, 176
World Confederation of Labor, 209
World Trade Organization, 194, 210
Writers Guild of America, 103, 134–35, 207

About the Authors

Vincent Mosco is Canada Research Chair in Communication and Society and Professor of Sociology, Queen's University, Canada. Professor Mosco graduated from Georgetown University in 1970 and received his Ph.D. in Sociology from Harvard University in 1975. He is the author of numerous books on the media, telecommunications, computers, and information technology. His most recent include *Knowledge Workers in the Information Society* (edited with Catherine McKercher and published by Lexington, 2007), *The Digital Sublime: Myth, Power, and Cyberspace* (MIT Press, 2004), which won the 2005 Olson Award for outstanding book in the field of rhetoric and cultural studies, and *Continental Order? Integrating North America for Cybercapitalism* (edited with Dan Schiller and published by Rowman and Littlefield, 2001). He is also completing a revised edition of his 1996 book *The Political Economy of Communication,* which is to be published by Sage in 2009.

Catherine McKercher is Professor of Journalism and Communication at Carleton University in Ottawa, Canada. She holds a B.A. in political science from Carleton (1974), a Master's of Journalism from Temple University (1985), and a Ph.D. in the Humanities from Concordia University (2000). Before joining the faculty at Carleton, she worked as a newspaper and wire service journalist in Ottawa, Toronto, and Kingston, Ontario, and in Washington, D.C. She is coauthor of *The Canadian Reporter: News Writing and Reporting* (Harcourt Brace 1994, 1998), Canada's leading journalism textbook. A third edition of that volume is in preparation. She is also the author of *Newsworkers Unite: Labor, Convergence and North American Newspapers* (Rowman and Littlefield, 2002). Her most recent book, coedited with Vincent Mosco, is *Knowledge Workers in the Information Society* (Lexington, 2007).

www.ingramcontent.com/pod-product-compliance
Lightning Source LLC
Chambersburg PA
CBHW051253200926
51612CB00001B/28
* 9 7 8 0 7 3 9 1 1 8 1 3 9 *